7-10-73

W9-DJJ-654

GARBAGE

By the author of

THE WONDERFUL WORLD OF
WOMEN'S WEAR DAILY

GARBAGE

*The History and Future of
Garbage in America*

KATIE KELLY

Saturday Review Press

NEW YORK

Published simultaneously in Canada by
Doubleday Canada Ltd., Toronto.

Library of Congress Catalog Number: 72-79053

ISBN 0-8415-0187-4

Saturday Review Press
380 Madison Avenue
New York, New York 10017

PRINTED IN THE UNITED STATES OF AMERICA

Design by Tere LoPrete

to HAG, Karen S.H., and Charles S.

CONTENTS

GARBAGE

I

An Introduction to Garbage

When five University of Oregon students and two teen-agers set out to climb Alaska's 20,320-foot Mount McKinley, they discovered that in addition to being on one of Nature's scenic wonders they were also on one of the highest and most unlikely garbage dumps. At the 17,200-foot level they found familiar reminders of civilization—junk discarded by other climbers and skiers: ski bindings, socks, even underwear, plus literally tons of paper being blown around the majestic mountain by the 100-m.p.h. winds that rake McKinley's frigid slopes. The climbers gave up their goal of reaching McKinley's peak. Instead, they began the monumental task of burning and breaking as much of the junk as they could, then backpacking what they could handle back down the slopes to camp. In all, the Magnificent Seven managed to stagger down the iced slopes with 380 pounds of litter. With just a little thought, and a modest amount of handy everyday equipment, people can produce garbage anytime, anywhere, and on an almost nonstop basis. Not that we are ready to rest on our laurels. Not at all. Just because the 17,200-foot level of a majestic mountain is cluttered with litter does not mean the battle is won.

Ever since our first tentative steps into the real world we have striven doggedly to maintain what we have come to consider America's own Manifest Destiny: supremacy, whether in warfare, celestial exploration, jazz, automobiles, or television situation comedies. We are truly a first-with-the-most kind of country. These are high levels of achievement that we have, over the decades, willingly set for ourselves.

There are those certain few areas about which some of us feel hesitant. Where some of us feel second-rate. We might have our Leonard Bernstein, but where is our Mozart? We might have our John Lindsay, but where is our Henry VIII? We might have our Ann-Margret, but where is our Lady Godiva? But there can be no doubt about our garbage. America produces over 360 million tons of garbage per year. No other country can begin to approach the amount of garbage generated by an alerted and dedicated populace such as ours. This figures out to approximately 10 pounds a day or 1.8 tons per year for each and every one of us. (India can scratch up only 200 pounds per year per person. Imagine how the Indians must feel. It is one thing to own the Taj Mahal; it is another to produce only 200 pounds of garbage a year.)

Three-hundred-and-sixty million tons of garbage. That is enough garbage to fill 5 million trailer trucks, which, if placed end to end, would stretch around the world twice. To shovel this pile of garbage out of harm's way costs American taxpayers $3.7 billion a year.

We spend only $130 million on urban transit, only $1

billion on urban renewal, only $1.5 billion on medical research, only $2.5 billion on food stamps and other nutrition programs.

Lest anyone think we have peaked in garbage production, that 360 million tons of garbage we hardworking Americans produce per year will look like a very small hill of beans indeed compared to what we will be doing in 1980, when that yearly mound will hit an impressive peak of 440 million tons of solid waste per year. That, folks, is a lot of garbage.

What must be noted, however, is that we are interested not only in quantity but in quality. What good would it do to have the biggest pile of garbage in the world if it were not the best pile of garbage? Within the challenge of quantity there is also the qualitative challenge. But fear not: Americans are up to it. Over our heads, as a matter of fact.

It would seem safe to say that America produces the best garbage in the world. First of all there is household waste, some 195 million tons of it. All the familiar stuff found in our garbage cans. Cereal boxes and potato peelings and Saks Fifth Avenue boxes and discarded copies of *The New York Times* and *Time* magazine and paperback books and dead leaves and dead cats and dogs and rats and canaries and old clothes and those broken TV sets and junked furniture. That's at home. If you gave at the office, add 45 million tons for computer cards, punch tapes, stationery, plastic-coated coffee cups, scratch paper, all the letters the secretary had to throw away and start over, press releases, announcements, invitations, memos, order blanks, invoices, reports. And from industry a solid donation of 110 million tons a year in very heavy garbage: construction and demolition debris—lumber, wiring, insulation, pipes, tubes, concrete, bricks, tile; figure in manufacturing residue that is left over from making all the things we throw away in the home and office, plus 10 million miscellaneous tons . . . well, it is easy to see where our 360 million tons come from. (This is not figuring in 550 million tons of agricultural waste per year to keep us in enriched bread and corn flakes, plus another 1.5 billion tons of animal wastes to keep us in Grade A Choice and boneless hams and barbecued chicken. Any nation that can produce—seemingly effortlessly—1.5 billion tons of cow droppings deserves some sort of credit. Then there is an additional 1.1 billion tons of mineral and mining waste.)

From *all* sources—household, commercial, industrial, agricultural, mining—America emerges with a garbage total of 3.5 billion tons of garbage per year. If that is too heavy to handle, try it this way: 7 trillion pounds. That is *almost* too much to handle. And indeed it is.

Having gotten our garbage production down to a science, if not an art form (well—art *does* follow life) we are thus able to come up with a varied and variegated collection that would make an underdeveloped nation cry. Each year we discard 7 million cars, 7.6 million TV sets (many in working condition), 62 billion cans, 43 billion glass containers, $500 million worth of plastic and cardboard packaging materials, 65 billion metal and plastic container tops. Think of the six-packs of beer that go into those statistics. And you thought you were just popping a few to get through the NFL season. The President's Science Advisory Committee says that each of us throws away 135 bottles, 250 cans, and 340 container caps every year.

Now, all of this is not to say that our garbage production is limited to home use. Home garbage production—a cottage industry of the most mammoth sorts—is, however, the major source of this country's garbage. Surprisingly, litter is a minuscule part of the whole garbage picture, only 1 percent of the total. But that amounts to 3.6 million tons, which by anybody's standard is a lot of litter. Litter is special because it's personal. Litter is something that can be done by everyone no matter how small, how old, how crippled, how infirm. Matrons do it. Executives do it. Gay, carefree children do it. Our senior citizens do their part, Sun City on the move, contributing one last item on our garbage agenda before making it to the big retirement home in the Sky. Paraplegics do it. Junkies, pushers, dentists. Cats and dogs.

Litter is, quite simply, anything that is discarded in the wrong place. Everything from Snickers wrappers (on the sidewalk) to dog droppings (in your neighbor's tulip bed) is litter. In a study on litter done by the National Academy of Sciences it was found that young people litter more than old, men more than women, country folks more than city folks, residents more than tourists. Littering, however, is for amateurs, the dilettantes living in the wonderful world of solid waste. In a pamphlet called "National Study of the Composition of Roadside Litter," it is revealed that only 1 percent of the American public can be classified as a litterer. Or as Elgin D. Sallee, the Corporate Director of Environmental Control for the American Can Company, figured, "Chronic litterers, those guilty of gross acts of littering such as throwing a beer container or newspaper on the roadside, park or lawn, are estimated to be about 1 percent of the general population." That 1 percent, however, counts out to 2.05 million. As Mr. Sallee also reminded us, "Packages don't litter, people do."

People make garbage. There are now about 205 million people in the United States, in a world population of approximately 3.6 billion. The current life expectancy in America is seventy-plus years. By the time the average American citizen has lived out the seventy years that are statistically allotted to him, he will have used 26 million gallons of water and 21,000 gallons of gasoline, eaten 10,000 pounds of meat, and drunk 14,000 quarts of milk. We are kept pretty busy consuming. We might have only just over

5 percent of the world's population but by very diligent application of the Protestant work ethic we manage to use 40 percent of the world's natural resources—air, water, minerals, timber. Not that we do not return anything for the generosity of the world's resources. In using up that 40 percent of the world's natural resources we create 30 percent of the world's pollution.

People always equal garbage, especially when they are highly motivated. On May 22, 1970, a Great Raft Race was held down the Chattahoochee River in Georgia to dramatize the severe pollution of that river and spur efforts—both private and public—to clean it up. Some three thousand people participated in the race, and another ten thousand watched the proceedings. According to Larry Patrick, a Georgia Tech senior who acted as the race coordinator, upon cleaning up after the race some twenty tons of freshly deposited debris were pulled out of the already overburdened river. Thousands of empty beer cans, at least eight cases of unpopped beer cans, plus an array of tennis shoes, undershirts, pants, and one bra.

We have not yet begun to fill in Grand Canyon. The valleys and gorges created by the skyscrapers of New York City must be a Mecca to the landfill freak. *Think* of the amount of garbage you could dump on Manhattan alone. (Actually, Manhattan is not the real, long-term answer: one year's supply of American garbage would cover Manhattan Island to a depth of thirteen feet. At that rate, the life expectancy of Manhattan as a landfill site would be only fifty to a hundred years, depending on how you wanted to fill it.)

With an annual onslaught of over 2 million tourists per year, Yellowstone National Park, for example, faces an annual garbage load of 5,700 tons. Picking it up costs $128,300 and disposing of it (three incinerators and two natural dumps) costs another $72,900 a year. The incinerators, naturally, pollute the air. The dumps—one at Trout Creek near the center of the park, one at Rabbit Creek in the West and just above Old Faithful—are using up just what their names imply: two natural creeks. One is staggered at the sight of those two small creeks, meandering through the scenic vistas of Yellowstone, filled with Kleenex packs, beer cans, sandwich bags, potato-chip bags, razor blades, coffee cans, plastic bags, prophylactics—all the leftovers of an affluent society. Additionally, the open dumps provide excellent scavenging areas for Yellowstone's bears, which in turn keeps them out of the natural ecocycle of things and gets them adjusted to human garbage. When the dump becomes too crowded, the bears meander through campsites to find food. Tourists get uptight, bears often panic and attack, and—well, you know the rest. Mutilated tourists and dead bears. All for the privilege of throwing away a half-eaten hot dog and a beer can.

Now we Americans are an independent lot, but a quick dig through our garbage cans would reveal certain truths. Approximately 50 to 60 percent of our municipal garbage is paper. Newspapers, *TV Guide*, shopping bags, letters, junk mail, advertising flyers, radical circulars, ecology booklets, wrapping paper, excess packaging around aspirin bottles and eggplants and crackers and banana cakes and Tootsie Rolls, plus paper napkins and party invitations and shoe boxes and dime-store bags. Some 10 percent of our garbage is lawn and garden waste—leaves and twigs and dead flowers. Only 9 percent is food waste like potato peelings and carrot tops and plate scrapings and stale bread and gravy leavings and tough meat and dried egg yolk. Then there are glass and ceramics (8.5 percent) and metal (7.5 percent), such as cat-food cans, soup cans, coffee cans, aluminum beer cans and disposable cake pans, and bimetal soft-drink cans. The remaining 6 percent is shared by rags, plastics, rubber and leather, and just plain dirt. (The Midwest Research Institute, an independent group working out of Kansas City, figures that of the total fully 14 percent represents packaging—paper, plastic, metal, glass. Thus packaging alone plays a great role in America's garbage. Forces are currently working around the clock to make sure we not only maintain our current level of packaging use but significantly increase it.)

The whole character of garbage has changed significantly in the last century. American garbage used to be composed mainly of food wastes, most of which we fed to farm animals. In the past few decades, however, that has changed significantly. Any country whose garbage content is 14 percent packaging is beginning to have something to boast about. Plastics, while just a piddling amount in the total picture, still represent 2 to 5 percent of the total—and growing. Plastics are well-nigh indestructible and nearly impossible to recycle, i.e., break down and rework for use again.

What are the reasons behind our growth in garbage? Technology, affluence, population, and attitude. Technology produces more, our affluence allows us to buy more, our population keeps growing, which keeps opening up new markets for technology. And as we grow, our attitudes change and with careless abandon we demand more—not necessarily better—products, which we discard with ease both legally (in our garbage cans) and illegally (on our streets and sidewalks).

Garbage collection is in itself one of the most hazardous occupations in the country. Garbage men are constantly losing eyes (flying glass) and arms (getting them caught in conveyor hoppers on garbage trucks), while catching diseases (from the flies and rodents that breed so happily in our garbage) and the devil (from dissatisfied housewives, some of whom have been known to take off after collection men, brandishing their garbage-can lids in a very threatening manner). And the American Medical Association figures that environmentally induced diseases, from things like air pollution, mercury-poisoned fish, noise pollution, and garbage, cost us $38 billion a year.

In the San Francisco Bay region there are seven major airports and thirteen smaller ones, with thirty-seven garbage disposal sites (sometimes referred to as dumps). Add to these factors nearly 55,000 sea gulls. Now the Metropolitan Oakland International Airport lies between two disposal sites, or dumps, and at the peak gull hours, some 4,000 birds have been caught winging it through the air space used by jet planes making their final landing approach. Where were the gulls going? To one of the dumps, located a mere mile away.

Given both the high quality and the high quantity of America's garbage, one would assume that garbage is given a very high priority in America. Not so. Garbage is one of the most misunderstood and, in most instances, ignored of our commitments. What do we do with our garbage? We throw it away. We operate on the principle of "out of sight, out of mind." Fully 85 percent of all our high-grade American garbage ends up in open dumps where it is prey to rats, mice, flies, fleas, mosquitoes, gas formations, water pollution, and bad mouthing and bad vibrations. The national Environmental Protection Agency figures that 94 percent of our dumps—or, as they are sometimes elegantly called, landfills—are inadequate.

We incinerate only 8 percent of our garbage, which is not a very good fate for it either. Most incinerators are so old or inefficient that big clumps of air pollution come belching out. Not only is this unsightly and unhealthy, but it tends to give garbage an even worse name: air pollution. Fully 75 percent of the incinerators were deemed inadequate by the U.S. Department of Health, Education, and Welfare in a recent report.

Since about 95 percent of our garbage is disposed of in dumps (including 2 percent in the ocean) and incinerators, this means that we are treating only about 5 percent of our garbage with anything approaching sensitivity and understanding, for it is this amount that is consigned to sanitary landfills. Now a sanitary landfill, to environmentalists, at least means that some thought has been given to the garbage. A sanitary landfill simply means the garbage is leveled off, covered with soil every night, and measured every so often to make sure it is not getting out of hand. Landfill is a good way to fill in gullies, valleys, gorges, marshes, sandy areas, and shorelines. Some people find this highly offensive, wondering why an irregular coastline, for example, has to be filled in and smoothed out by garbage. Or why the natural ecology of a marshland—the plants, birds, insects, fish—has to be destroyed so society will have a place to dump its pop-tops and cub-size cans. It is worth pondering. For those who find beauty in gullies and gorges and craggy spots—those irregularities and momentary diversions that are considered a feast for weary eyeballs—garbage in the gorges is hardly gorgeous.

And so it goes. But we are a country of contradictions. On the one hand we expend time, energy, and effort building up the world's most impressive supply of garbage. Then we consign it to unsightly dumps, burn it up to form air pollution, or use it to fill in what some folks consider scenic wonders. Mostly, however, we do one of two things. We forget about it. Or we bitch about it. It smells, it's ugly, it's depressing. Garbage, truly, is one of our most misunderstood majorities. A shabby apartment is considered a "dump"; our prisons, schools, and armed forces are considered "dumping grounds" for misfits and misbreeds. A person being maligned is said to be "dumped on." A defecating dog is said to be "taking a dump." A handy and current pejorative is to call something disagreed with "garbage."

When someone is losing weight rapidly they are said to be "wasting away." To kill someone, in junkie vernacular, is to "waste" them.

Fortunately, we are not working alone in our race against time to produce more and better garbage. Where would the consumer be without American know-how: technology, advertising, affluence, population. Indeed, garbage is the effluence of affluence. We produce, because that's the American way. And lest we fall behind in our consumption, we have Madison Avenue encouraging us to buy. We are Super Consumers in a Super Market of goods. Not only do we buy what we need, but we are encouraged to buy what we do not need. And everything is made so convenient, so colorful—who can resist? General Motors and General Foods and Colonel Sanders are there to lead us. Keeping the home fires burning are Betty Crocker and Aunt Jemima and Uncle Ben.

Yes, this is a serious problem—our garbage. One worthy of our attention and our concern. A problem well worth our consideration, for it involves each and every one of us, every single day of our lives, raising questions that must be answered: can a nation of dedicated consumers raise their already high level of consumption even higher, thus creating a garbage supremacy no other country could begin to hope to meet, much less match? Can we keep the widening gyre of affluence-production-promotion-consumption-effluence spinning around indefinitely? Or will we one day be confronted with the spoilage of our labors: a garbage gap? A stockpile of garbage that we, the victors wallowing in our own spoils, cannot dispose of?

II

Garbage Through the Ages

In the beginning there was Garbage.

There can be little doubt about it: Eve was the first litter-bug. Although there is little proof of it, it is doubtful that she—given her panicky and probably paranoid state of mind at the time—ate the entire apple. One bite was all she had to take, and when the sound and the fury arrived one can hardly suppose Eve stood there calmly munching on her Golden Delicious while original sin was being promulgated, the gates of Heaven were swinging shut, and the fires of hell were being kindled down below. No, calm such as this was probably not the order of the day in the Garden of Eden while fire and brimstone were bubbling away in the celestial caldrons. Hence, once that fateful bite was taken, it seems safe to assume Eve tossed her unfinished apple to the ground and went on to other things.

As for the apple, there it lay letting newly formed nature take it course. Yes, garbage was simpler in the good old days. One can assume that natural decomposition probably set in while worms and ants and slugs and such completed the cycle and the apple went its natural way: into the food chain. But consider the consequences of that first litter bit. People too often tend to concentrate on the initial act—the bite from the apple—rather than taking into account the entire chain of events. For in addition to the garbage created—the core hitting the ground—that initial act of use and subsequent disposal set off a chain of events that the Holy Mother Church daily traces directly through history and right down to the present day. Pain in childbirth, wars, pestilence, sinus congestion, family feuds, aches and pains: all of these are direct consequences of that rash act of Eve's. Today it is only a fifty-dollar fine. Careless actions and reactions are fraught with dangers. That first litter bit was, indeed, the Eve of our destruction.

It is generally conceded by the historical garbage experts (there are few of them, so academic dissension on these points is rare if not nonexistent) that Eve's natural reaction —the tossing away of the apple—was probably the order of the day for decades, centuries, eons, and ages. Animals were killed and every conceivable part was used in some way: the meat was eaten, the fur was used for clothing, bones for weapons and, later, crude utensils. What little was not used was tossed away on the spot, thus creating the first "landfill."

This trend lasted well into the nineteenth century: User-ism. What the cavemen did was simply make sure they brought in only what they needed and used it all up. If one brontosaurus was all they needed to make it through the winter, one brontosaurus it was. Why indulge in overkill just to impress the neighbors? If one hairy mammoth pro-vided enough fur for their clothing and a little left over for

throw pillows—fine. And why decimate the whole berry
bush when the recipe only called for a handful? Thus very
little was created in the way of "garbage," i.e., something
that is left over for discarding. And naturally they did not
have to worry about discarding the Sunday *Times* or spending any of their precious hunting time stomping cans, separating their bimetals from their all-aluminums, and then
hauling the whole lot off to the neighborhood recycling
center, operated by a lot of prehistoric do-gooders. Life was
not only neater, but simpler.

It was left to fifth-century B.C. Greece to begin working
seriously on the idea of town dumps. These unsanitary
landfills were kept in excellent repair by the constant and
imaginative use by the citizens of the appropriate towns as
garbage and refuse of all sorts made its way to these garbage dumps. Among the usual deposits normally made at
town dumps—foodstuffs, fecal matter, potsherds—Atheni-
ans also disposed of unwanted babies. Mistakes of nature
and bastard sons. The kids would be spirited out to the
town dump *à la* Oedipus and left there to biodegrade into
some high-grade compost.

Well, as urbanologist Lewis Mumford once observed of
the noble Athenian, "If he did not see the dirt around him,
it was because beauty held his eye and charmed his ear. In
Athens at least the muses had a home." True enough. Often
in the town dump.

Greece was also working out an effective compositing
plan. Knossos, an ancient capital of Crete that flourished
some thirty-five centuries ago, had compost pits at Kou-
loure for depositing of their organic wastes.

Garbage men were organized for the first time in the
Roman Empire. It became a familiar sight, one that has
been nobly enacted and reenacted with minor variations
down through history to this very day: the garbage man
making his appointed rounds. A wagon, pulled by a horse,

accompanied by the world's first sanitation men, wearing togas. As usual, the streets were the logical receptacle for the garbage and—since neither paper bags nor plastic disposal bags had been invented—garbage was tossed in its original condition (loose) into the streets. Legions of Roman sanitation men picked their way down the various vias, scooping the stuff into their wagons. From time to time certain centrally located spots would catch someone's eye and handy neighborhood dumps would be created. Then, as today, dumps were not the most popular form of landscape architecture. Then, as today, they were unsightly, they smelled, and all sorts of nasty things crawled around on them—rats, flies, mosquitoes, and an assortment of Roman nuts and kooks. Hence Ancient Rome was often emblazed with signs reading: "Take your refuse farther out or you will be fined," replete with arrows pointing away from town.

Yes, there is something to be learned from history. Rome at the time of the Caesars (27 B.C.–A.D. 410) was famous for its aqueducts, which would carry fresh cold water hundreds of miles down from the mountains and into the city. Roman engineering feats such as these were dazzling. The Pantheon. The Coliseum. The Temple of Venus. But as far as garbage disposal, it was not a priority item with the Romans. It showed. And smelled. Carcasses—human and otherwise—were normally disposed of in open pits at the edge of town. This was known to send up a most god-awful stench, particularly during any Roman spring. And summer. And particularly when the wind blew toward town instead of away from it.

Not that Rome did not have its very own particular, unique carcass problems, for on any given day bodies from gladiatorial encounters would often hit as high as five thousand animals and hundreds of human beings. Then, as today, contact sport was not without its problems. Even an

amateur, or a person with only a passing interest in garbage, can well appreciate the problems of disposing of five
thousand animals whose numbers included creatures the
likes of elephants and camels. Not to mention all those used
lions.

Although it was illegal to dispose of fecal matter in carts
and open pits, the sanitation subcommittee of the Roman
Senate made no laws about the disposal of just about everything else in that manner. Including those elephants and
camels and all those human losers in the arena. Rather
indiscriminate dumping, you might say. (Just to make up
for it—or to recognize the problem, perhaps—Rome had an
inordinate amount of activity in the way of altars and
shrines dedicated to the Goddess of Fever. One busy lady,
she. Alas, it did not always work: Rome was plagued, as it
were, in 23 B.C. and A.D. 65 and A.D. 79 and A.D. 162.) These
pits retained their characteristic redolence down through
history, for when archaeologist Rodolfo Lanciani did a little digging in some Roman pits he was "obliged to relieve
my gang of workmen from time to time, because the stench
from that putrid mound, turned up after a lapse of twenty
centuries, was unbearable even for men inured to every
kind of hardship, as were my excavators."

The idea of collecting and disposing of garbage in any
sort of systematic and sanitary way, however, really did not
catch on until well into the nineteenth century. Those
early practices in Greece and in Rome were more the exception than the golden rule and fell into misuse through
the Dark Ages and even through the Renaissance. In fact,
littering was the most common form of garbage disposal
until the 1700s and 1800s. Until then, citizens just threw
their garbage into the unpaved streets and roadways. (Today that is often referred to as the "airmail" method of
disposal, deriving from certain city dwellers' quaint habits
of throwing their garbage out their windows into streets,

sidewalks, lots, and yards.) Ancient garbage, however, would be absorbed into the roadway, where it would mingle with the human excrement and animal droppings that were also there. All would then decompose naturally and organically. Mumford cites that even in the Golden Age of Greece the "narrow, crooked streets of Athens were heaped with refuse." Later, local ordinances were enacted to forbid the dumping of garbage in the streets but, like antilitter laws of today, they were sissy laws. Nobody paid much attention to them. When the cops were called they gave familiar excuses. We are busy. We have more important things to do. Peloponnese is attacking—you expect us to shovel up grape leaves and lamb bones? Times have not changed all that much.

For all those who believe in Hollywood technicolor spectaculars, forget it. For one thing, those medieval castles were damp, dank things without the glories of central heating, much less indoor plumbing. Dining rooms were rarely covered with rich carpeting. And those tapestries on the walls? To keep the cold stone walls covered and to hold down the drafts. And most of the time the floors were covered with straw (to hold down the dank and the damp), into which went an assortment of droppings for and by the castle dogs. This also meant the floors did not have to be swept constantly. Or at all. Food scraps and dogs being what they are, however, things got pretty ripe in the main dining room every so often. When that happened, the floors would be scooped and a new layer of straw would be spread about. Rather than sweep majestically about his kingly halls, what Henry VIII really did was pick his way through layers of straw loaded with doggy droppings and food scraps and all the little things that go bump! in the night while they scurry through the scraps. (One theory says that Christianity and greed were less to blame for the warring feuds that broke out during the Middle Ages than the fact

that the lords couldn't stand hanging around their baronial
—and kingly—estates for long periods of time. Between the
plumbing—nonexistent—and the garbage—too existent—
things would just get to be too much. So King A would
provoke King B, and off they'd go.)

Life was freer if not more fetid way back when. But
tentative efforts toward refining and regulating garbage
production and disposal were being made. In 1388 the Eng-
lish Parliament passed an act forbidding the dumping of
filth and garbage into ditches, rivers, and waters. By the
sixteenth century in London, for example, it was ruled that
"no man shall bury any dung within the liberties of the
city" nor "carry any ordure till after nine o'clock," when,
it was figured, all good souls would be in bed and asleep and
not hauling their garbage out to dump it illegally in the
rivers.

The late 1500s, however, can be considered a high point
in the history of garbage, for it was then that an entirely
new form of solid waste was being introduced to western
culture. Not only was the modern water closet invented
(Sir John Harington, 1596) but shortly thereafter came a
notable Chinese import: toilet paper. The two of these together moved the cause of civilization ahead centuries.
From there it was but a short step to Hudson "Together"
trays made of colorful, nonbiodegradable, breakable plastic, to colorful mix 'n' match ('n' pollute) bathroom tissues
and toilet paper.

Versailles, believed by some to be the first Hilton Hotel
ever constructed, was Louis XIV's modest country retreat
just outside Paris. Versailles may have been opulent to the
point of no return, but it was deficient in a few areas: for
one thing, the plumbing. The stench got so bad that using
cologne (which was actually invented in Germany) became
a very French thing to do. Legend has it that some *nouveau
riche* nobles, hovering just the other side of social respecta-

bility, were out to copy Versailles court life right down to the last minute and intimate detail. And, in order to dupli-cate Versailles' own unique smell, these noblemen would pay volunteers a nominal sum to come dump organic deposits right near their manor house. Versailles' plumb-ing deficiencies extended, of course, to the garbage dis-posal. So, much like his peasant subjects, the *Roi Soleil* et al. slopped a lot of pigs with his garbage and spread the rest of it here and there to let it do what came naturally (first of which was to smell like hell). Of course, Louis, like the lowest commoner, was not encumbered by garbage made up of 50 to 60 percent paper, 16 percent bottles and cans, and 2 to 5 percent plastics.

When the Second Republic in France was overthrown by Napoleon III in 1852, French town planning got a big boost from Baron Haussmann. The baron was the first to establish cemeteries on the outskirts of a modern city, thus providing a disposal place for Parisian loved ones (deceased) in addition to providing "a green ring of mortu-ary suburbs and parks around the growing city." It beats the Roman approach to corpses six ways to Sunday.

That first Thanksgiving which was so joyously cele-brated by the Pilgrims and the Indians in 1621 is a prime example of both human and ecological cooperation. For one thing, the problem that plagues Thanksgiving din-ners today was not even thought of back in 1621: leftovers. There were none. Food was scarce, so everything was eaten. Any scraps or droppings or leavings were fed to the animals, bones included. There were no Shake 'n Bake bags to dispose of. No ready-bake aluminum pie tins to get rid of. No vegetables to be cooked in their own butter sauce inside a sturdy plastic bag. No beer cans and Hawaiian Punch cans and V-8 juice cans to be gathered and disposed of. No prepared stuffing bags to contend with. No aerosol whipped-cream canisters to worry about. The gang just gathered, fixed the food, ate it, belched, and went home.

The American Indian was perhaps America's first and finest example of a Use-ful society. Porcupine quills and seashells and feathers and teeth—all the things that were left over from foodstuffs and everyday clothing—were artistically incorporated into ceremonial dress. Meat was eaten, the fur and hides were fashioned into clothing and homes and bedrolls and weapons. When the first settlers arrived, however, they began teaching anyone who would witness their activities the glories of garbage production. Although on a limited scale, early Americans did produce their fair share, given the obvious limitations under which they were working. If they could not abandon cars in the streets, they could certainly abandon wagons and homes and household belongings that they had grown tired of or for which they had better replacements. In fact, that old pioneer spirit of "Go West, young man" was in part due to early efforts at pollution: when one home site would have a superabundance of pollution around it—utensils, equipment, rags, excrement, animal wastes, and what have you —Ma and Pa would simply load up the kids and move on to greener pastures. Thus the very westward history of our great nation can, in part, be traced to pioneering efforts in pollution production.

Those great Westward Ho! treks that became such a popular form of diversion during the late 1800s here in America were, however, hints of future happenings. For the westward migrations saw the establishment of the first recycling centers this country has ever known. Molly's rosewood piano, too heavy to haul over Pike's Peak: leave it by the road. Sarah's trunk all full of her hopes and dreams for the future good life somewhere over the rainbow of the Rockies, too heavy for the hungry ox team to pull over the virgin ruts of the Oregon Trail: leave it behind somewhere along the Platte River Valley in Nebraska. Trunks, brass beds, pianos, tables. All the lovely additives to the good life, but hardly necessities when you are caught in the crossfire between Custer and the Sioux. Hardly necessities when a breathless white-hot dust storm blows up out of Kansas or when a stunning white-cold blizzard whistles down out of the Dakotas. Leave those hopes and dreams behind and pray someone else has both the good fortune and the room to make use of them.

Some early Americans, however, were aware of what they were doing, and their awareness manifested itself in an appreciation of their garbage. Thomas Jefferson at Monticello, his magnificent home near Charlottesville, Virginia, devised an intricate and elaborate underground conveyance system. Garbage and scraps and ashes were loaded onto buckets, which were carried by a pulley system some distance from the house for disposal. It was the first pneumatic garbage disposal system installed in the continental United States.

Benjamin Franklin has long been regarded as one of our most productive citizens. The lightning rod and the Franklin stove and bifocals . . . but how many remember Franklin for his innovative garbage techniques? Franklin established the first systematic garbage pickup service for Philadelphia.

In 1792 Philadelphia's population was 60,000. Slaves carrying loads of garbage on their heads headed for the Delaware River, where they waded out and deposited their garbage downstream from the city, all according to a plan mapped out by Franklin. Franklin might have been born in Boston, but he left his mark on Philadelphia.

By the 1800s garbage and the growth of cities were inexorably linked. The Siamese twins of society, garbage and growth. Where would one be without the other? Growth would indeed be a shallow and meaningless commodity if there were not leavings and droppings as reminders of civilization's progress. Once it was determined that one depended very much on the other—garbage and growth— a bond of mutual understanding and mutual cooperation was forged. A bond of garbage.

As Dr. Charles Gelb (associate professor of History and director of Urban History Section of the State Historical Society of Wisconsin) put it, "In sections of all cities rubbish and garbage was piled in gutters in the streets or left to accumulate in back yards and alleys." Hogs, geese, and dogs scavenged in the streets. In Charleston, West Virginia, they—and the friendly neighborhood vultures— were even protected by law for the scavenging duties they performed for the citizenry. Bands of rats roamed under the sidewalks and fairly ruled the downtown portions of larger cities. (New York finally became aroused when, in 1860, a newborn baby was devoured by a very bold rat who operated out of Bellevue Hospital.)

A social observer, John H. Griscom, who did a report in 1854 on sanitary conditions among the working classes of New York City, wrote: "Every resident who takes a stroll into the country can testify to the difference between the atmosphere of the two situations. The contrast of our outdoor atmosphere, loaded with the animal and vegetable exhalations of our streets, yards, sinks and cellars—and the air of the mountains, rivers, and grassy plains, needs no epicurean lungs to detect it."

In the 1890s Washington, D.C., used to load its garbage on barges and float the stuff down the Potomac to a point just south of Alexandria, Virginia, for disposal. Now Alexandrians are a patient lot, full of sympathy and understanding, but this noxious practice was simply too much. Taking a cue from their Boston ancestors—who took both matters and crates of tea into their own hands—a number of infuriated Alexandrians were known to commandeer various garbage skows and sink them in the Potomac. Even today, the bed of the Potomac is littered with the remains of this weird warfare that raged—not up and down the Potomac, only near Alexandria—during the last decades of the 1800s.

New York got a leg up on her garbage problem with the introduction of Belgian paving stones to the city in the 1800s. Not only did this tend to make the carriage ride a little easier and a little quicker, but our street garbage did not sink into oblivion so quickly after that point. Small American towns still had the most effective disposal: automatic composting and on-the-spot pig feeding. Cities, however, were experiencing new difficulties as they grew larger and larger.

By the late 1800s New York's garbage was still growing. George Strong, an intelligent citizen of the time, was a prolific diary-keeper. "Such a ride uptown!" he wrote on June 19, 1852. "Such scalding dashes of sunshine coming in on both sides of the choky, hot railroad car. . . . Then the feast of fat things that come reeking under one's nose at each special puddle of festering filth that Center Street provided in its reeking, fermenting, putrefying, pestilential gutter! I thought I should have died of the stink, rage, and headache before I got to Twenty-first Street." Believe it or not, he loved Greater Gotham, despite its mud, "the consistency of molasses," and its overall "civic filthiness."

American literature is filled with bypassing references to our growing garbage problem. In *The Jungle,* published in 1906, Upton Sinclair described the area in Chicago to which his Lithuanian immigrant Jurgis Rudkus has come to find liberty and pursue happiness. It was a grim, barren place full of bare spots "grown up with dingy, yellow weeds, hiding innumerable tomato cans. . . . There were no pavements—there were mountains and valleys and rivers, gullies and ditches, and great hollows full of stinking green water." There was "the strange, fetid odor which assailed one's nostrils, a ghastly odor, of all the dead things of the universe. It impelled the visitor to questions—and then the residents would explain, quietly, that all this was 'made' land, and that it had been 'made' by using it as a dumping ground for the city garbage." Good old Jurgis Rudkus leaves his native Lithuania with its forests and fields and wildflowers and wide open spaces and comes to America, the land of promise. What does he find? He finds himself living on a landfill, an unsanitary landfill at that. "After a few years the unpleasant effect of this would pass away, it was said; but meantime, in hot weather—and especially when it rained—the flies were apt to be annoying. Was it not unhealthful? the stranger would ask, and the residents would answer, 'Perhaps; but there is no telling.'"

By 1874 the world's first incinerator—called a "crematory"—was built and began operating in England. Not to be outdone by the motherland, America had one of her first incinerators by 1885. In 1896 the Germans had mobilized their technology, and the first incinerator was built for the Continent.

Appropriately enough, America's first incinerators were built on Army posts. Garbage and the Army, inexorably linked, would thus march through history into greatness together. First in peace, first in war, first in garbage. *Semper fidelis ex garbage.*

The race was on. Incineration fired the fickle public's imagination and became the hot issue of the day. At that time, nothing was too good for our garbage, especially new-fangled inventions like incinerators. Allegheny, Pennsylvania, used the first municipal incinerator in 1885, followed closely by Pittsburgh and Des Moines in 1887 and Yonkers, New York, and Elwood, Indiana, in 1893. America was on her way, at least for a time, in providing the very best that technology could build and money could buy for her garbage.

In those days we separated garbage, like Mom doing the laundry: wet stuff here, dry stuff over here, salvageable materials over there. Then we burned it or buried it or sold it or reused it. Little cottage industries sprang up: the old-clothes man, the tin-can man, the ragman, the paper man, the compost-heap man. Molly Malone had nothing on the secondhand man. Cockles and mussels be damned—the junkman was an industrial spin-off to be reckoned with. Every city and town and village and hamlet the length and breadth of this great country of ours had one.

But then grim days descended upon us and our garbage. Gone were the good old days of cooperation between cottage and corporation. Gone were the good old days when there was a balance of trade between what we produced and what we used. When we produced only what we needed, and needed all we used. Soon America found herself producing and purchasing more—using up and utilizing less. We had cars left over. We had refrigerators left over. We had black-and-white television sets left over. We had packaging left over. We had foodstuffs left over. We had newspapers left over.

We are in dark days indeed. Like a cold moldy shroud, our garbage has piled up around us. Suddenly we have found ourselves confronted with the monster of our own making: an ever-increasing, never-diminishing pile of garbage. It grows higher and higher, a tower of garbage, threatening to fall down on us. If today we are still searching for the mythic civilization of Atlantis, will future generations seek fruitlessly for twentieth-century America? Will they send teams of archaeologists to dig through the garbage pit of our cities? Hey—here's a wide-track Pontiac. Look—a Quasar TV. My God—a real genuine imitation American town: DISNEYLAND!

III

Steve McQueen
Is an Ecological Disaster

Make no mistake about it: we have built the better mouse-trap. It is all things to all people. And here we are, caught in it, chasing after the elusive butterfly of progress. It is a Sisyphean effort, to say the least. The more we chase, the more there is to chase. Just when we think we've netted the butterfly, out comes a new model. Just when we've made the final payment on our Master Charge, they up and change the front end or the color or the fine tuning.

We are caught in a vicious circle. First there is technology, hell-bent on turning out all of everything we, the market, will bear. And our technology is a truly wondrous thing. It has given us A-bombs and H-bombs and put us on the Moon. It has given us an automobile population of 85,000,000. We have nearly 200,000 airplanes flying nearly 4 billion passenger miles per year. We are a nation on the move full of mobile homes and dune buggies and motorcycles and snowmobiles. We have flip-top cans, cucumbers sealed in plastic, individual pudding cups, and that truly ethnic all-American institution, the TV dinner. We Roast and Boast and we Shake and Bake. We are a Light n' Lively, Brisk 'n Bouncy people who have instant coffee and instant tea and instant pudding and instant soup and instant cake. We have make-believe orange juice and make-believe butter and make-believe milk.

Then there is Madison Avenue making sure we notice it and buy it. Each day, says the Television Bureau of Advertising, we are confronted with 560 advertising messages from newspapers, television, radio, bus placards, subways, billboards. Everywhere we look, our landscape and our futures are brightened by nervy ads telling us our armpits will stay drier, our breath sweeter, our scalp tinglier, and our weekends a lot more fun if we buy Brands X, Y, and Z not to mention A to W.

Then there is affluence. A little extra change in the pockets. Long-term loans. Credit cards and charge accounts. Buy now, pay later. Fly now, pay later. Buy, buy, buy, and it's only ten dollars a month for the rest of your life. Everything from birth to death, with a retirement home in Florida in between, can be gotten on some sort of credit plan.

Finally there is population, which is our own personal numbers game. We are doubling up at a faster and faster rate. It took 180 years—from 1650 to 1830—for the world population to double itself. In the next 100 years it doubled itself again. Now we face another doubling by 1975 if we continue doing what we are doing.

Technology, advertising, affluence, population.

And they all add up to garbage.

As scientist Barry Commoner points out, our technology has faulted us for one very important reason: it deals with only one problem at a time. It deals with "how to do it" and forgets about "what to do with it" after it is made and discarded. It does not deal, at the time of conception, with what to do with nonbiodegradable plastic. It does not deal with detergents. "No one troubles over the fate of the material when it has outlived its usefulness," says Dr. Commoner, "and is intruded into the environment."

Plastic is a perfect example of how our supratechnology has failed us. Since World War II our production and comsumption of plastics has skyrocketed. Most plastic products are petroleum-based, which means (1) more markets for oil, which, of course, means (2) more drilling in our oceans and Alaskan wilderness areas and, of course, (3) more dangers from oil spills and oil slicks. Beyond that, plastic—as a sturdy man-made substance—does not break down naturally. Throw it away, and plastic is forever.

For we do indeed live in a circle. Or, rather, we used to. And we still should if we have any hope of surviving the conflagration we are setting around ourselves. Until 1941 the discards of our use were, for the most part, organic. Biodegradable. When we discarded them they went back into the natural cycle of things to be broken down and used again. But with the advent of World War II and the rise of synthetics—particularly plastics—we entered a technological age that produced nonorganic materials that did not go back into the cycle. We broke out of our comfortable and workable circle, using nylon instead of cotton and plastic instead of paper. Dynel instead of real hair.

"We are in an environmental crisis," Commoner says, "because the means by which we use the ecosphere to produce wealth are destructive of the ecosphere itself. The present system of production is self-destructive; the present course of human civilization is suicidal. Now that the bill for the environmental debt has been presented, our options have become reduced to two: either the rational, social organization of the use and distribution of the earth's resources, or a new barbarism."

Granted, American technology has freed us from the shackles of manual labor. We have washing machines instead of washboards. Refrigerators instead of iceboxes. American technology also saves lives. Witness open-heart surgery and the strides made in nearly eliminating communicable diseases from our lives. A kidney dialysis machine looks as complicated as a computer—indeed it is— and American technology brought it into being to save lives every year. The list of timesaving and lifesaving innovations brought about by American technology is endless and encouraging.

It is also depressing and unnerving. If America is anything it is a land of very big business. If American big business is anything, it is (1) competitive and (2) self-serving. The profit motive ranks high, and that means quick sales and ever higher sales to keep profits high to keep stockholders happy.

Periodically, the American consumer wonders why things don't work better. Why that car he bought last year for $3,000 is today, only one year and less than 10,000 miles later, only worth $2,000. Why that Polaroid camera, the newest of its kind last year, now boasts a newer model that comes with still another button or buzzer or feature that makes his old 1972 model look old-fashioned, obsolete. Why that electric hair dryer from last year was not equipped with the steam feature of this year's model. Why last year's refrigerator had those old-fashioned ice-cube trays while this year's model silently turns out an endless supply of ice cubes. While you eat, while you sleep, while you bicker and dicker and fornicate, this year's model stands staunch and silent in the kitchen, in the bathroom, in the garage, becoming obsolete.

American technology has perfected obsolescence. Make no mistake about it: it does indeed exist. First, American business plans things that do not last and, God forbid something should last forever, second, it makes sure they become outmoded, outdated, out of style.

"It is very definitely possible to design products that will last a great deal longer than they do," says Morris Kaplan, director of the Consumers Union, a nonprofit testing organization that continually subjects the fruits of industry's labors to close scrutiny. "Cases of built-in obsolescence are common," charged Joseph Martin of the Federal Trade Commission, in a speech at Northwestern University in 1971. "This planned obsolescence is accomplished in several ways—by failing to provide a source of spare parts for the reasonable life of the product. By making frequent style or manufacturing changes so that the user feels he must turn in his old model for one which gives him better performance. By including certain components made of materials which have a shorter life than the reasonable life expectancy of the product itself."

Early in 1971 the Federal Trade Commission in Washington began digging into the whole question of "planned obsolescence" and "replacement parts obsolescence," which they called those "situations in which the possibly unjustifiable absence of replacement parts may have the effect of shortening the useful life of products." The FTC, moving with the normal ponderous speed of most other federal bureaucracies, figured it wouldn't have anything ready until well into 1972. Most consumer cynics figured that where there's smoke there's fire, and that the FTC was just a few decades late in heading out to the blaze with a cup of water.

Planned obsolescence equals garbage. Tail fins on the

new car: junk the old. New color-coordinated kitchen utensils: throw the old ones away. New tuning controls or color
dials or whatever on the TV set: get a new one. Out goes
the old, in comes the new. Thousands of appliances—many
of them in working condition—hit the streets every year
for the garbage man to haul away. What was once a usable
commodity in our homes and our lives—junk.

As if "planned obsolescence" is not enough for the
American consumer to contend with, there is slipshod
workmanship. Buttons that fall off coats as soon as you
unwrap them and legs that refuse to stay on tables and
refrigerators whose plastic accouterments break off as soon
as you plug them in are small stuff. Consider the headline
"6.7 Million Cars Face GM Recall." GM hustled around to
recall that many Chevrolet Novas and Camaros along with
some Chevrolet and GMC light trucks spanning the years
1965–1969. The recall involved GM motor mounts, which,
the government insisted, would break, causing the engine
to rise up a few inches from the engine compartment.
When this happened, it would jam the throttle open and
snap the power brake lines. This would, quite simply, leave
the car running wide open—with no brakes. GM, of
course, insisted nothing was wrong and suggested the
driver "turn off the ignition." (Then what? Coast to a stop?
Stick your left foot out the door and drag it in the street?)

That monumental 6.7 million car recall was rivaled only
by GM's 1972 recall of 4.9 million of its vehicles for faulty
exhaust systems, which already had led to carbon monoxide deaths.

Nowhere is the theory of planned obsolescence more
accepted and more widespread than in the fashion industry. "We operate on the theory of planned obsolescence,"
explains James Brady, former publisher of *Harper's Bazaar*.
Up go the hems, out go the long skirts. Down go the hems,
in come the new lengths. The gypsy look, the classic look,

the Chinese look, the ethnic look; one year it's pink, the next it's basic black. Meanwhile, the amount of garbage mounts. Unless, of course, there is a cousin willing to wear last year's peasant look. Or a person is into rug braiding or quilting as a way of recycling outmoded clothes. Every year $55 billion is spent on clothing in the USA. Surely it is not all replacement of worn-out skirts and sweaters.

Then there is the automobile, which has been called "the perfect pollution machine." Look at the problems of car ownership today: the manufacturing of an automobile depletes our natural resources. Then in eight years—today's life span for an average car—it becomes a piece of scrap metal to be discarded at $10 to $16 per ton. And the trend these days seems to be toward wholesale junking of cars. There are 85 million private cars on the streets and highways of this country today. Seven million are abandoned each year. The automobile is the biggest disposable we have. The biggest piece of litter. Even Lady Bird, sitting under the Lyndon tree, could not rid America of auto pollution during the heyday of her Keep America Beautiful campaign. And there are those, across the length and breadth of this great land, who feel that if Lady Bird could not drive the problem from our shores, nobody can. It is that big a problem.

The effluence of our affluence. They rust, they deteriorate. They are objects of ridicule as sportsmen use their taillights for target practice. Ebullient youngsters smash their windshields to watch the safety glass shatter. Citizen groups demand that great high fences be built around them. Sanitation men who cannot clean streets around them complain and bitch and moan and groan and eventually have to haul them away.

(Actually not all abandoned autos are sources of despair. The urban junkie when confronted with this brand of urban junk can strip it down and recycle it into the vast auto underground—the joys of interchangeable parts. The urban junkie car is an on-site auto mechanics course with a Mercury Cougar ready for recycling.)

Add to those 85 million disposable cars 2.5 million motorcycles, 16.5 million trucks, and dune buggies, and—the latest treat—the snowmobile. That is a lot of combusting going on about the country.

Not that America has not thought through her dedication and commitment to the horseless carriage. Indeed, we take our technology seriously. Anything worth spending in the neighborhood of $200 billion on is worth something —especially our concern. Look at the garbage. First of all, those 7 million junked and abandoned cars each year, which is a lot of scrap metal cluttering up our roads, highways, and streets. Then there are some 200 million tires that are discarded each and every year. To burn them pollutes the air. They rot at a very slow rate as they sit there along the road or in a field or in a vacant lot.

But more depressing are the accident statistics. In a very real and horrifying sense, the people who die in automobile accidents—from Motown's supercars with all that unnecessary horsepower—are a form of solid waste. It is wasteful indeed to think of Christmas and Thanksgiving and Labor Day and the Fourth of July and Memorial Day in terms of statistics. In 1950 34,063 people died in automobile accidents. In 1960 that figure was up to 38,137. By 1967 it had hit 52,924 and in 1971 a total of 55,000 people were killed in automobile accidents in the United States.[1] Fifty-five thousand people. Think of the human tragedies locked within those cold, unfeeling numbers. That statistic is fairly throbbing with motherless children, lovers, old people. People.

---

1. Additionally, some 2 million people suffered disabling injuries in 1971 owing to auto accidents. No wonder Howard Hughes stays indoors.

Not that industry ignores its responsibilities altogether. Not at all. According to one of Ralph Nader's reports, General Motors installed seat belts costing $6 and passed the safety factor on to the consumer for prices ranging from $23 upward to $32. Good old American free enterprise.

There are those among us who pride themselves on appreciating the finer things of life. Sunsets, the Rolling Stones, forests, sweet-smelling breath, Sara Lee cakes, fresh water, neat armpits, Steve McQueen. Yes, Steve McQueen, the all-American Boy and his all-American hobby: Steve likes to get on his cycle and roar around the desert for relaxation. Or hop in his dune buggy and churn up the silica just to get the kinks out. But how many have stopped to think of Steve McQueen as symbolic of ecological disaster. Yes. It is painful when one's hopes and Great American Dreams go up in a cloud of exhaust smoke. That motorcycle—and thousands like it—are becoming the bane of the existence of the California desert, all 16 million acres of it.

The Mojave and the Colorado—which sweep down from Death Valley in the north to Mexico in the south—are a dun-colored parade of dunes. The desert is a wild and wonderful place, full of nature's sound and fury: small animals who never drink water, reptiles who shrink off and seem to disappear into the sand, exotic cactus plants urging their own special brand of life out of the simple hostility of the desert. All the flora and fauna. And all the beer cans, bologna sandwich leavings, plastic Baggies, broken nonreturnable bottles, Pampers, charcoal briquets, tires, and noise, and garbage. Eleven million people visit the desert yearly, most of them from the southern California sprawl that is yearly inching eastward, threatening the desert in a wholesale way. The Department of Interior, in a depressing report on the desert in 1970, said: "Unless we act soon the desert faces damage and destruction because of people pollution."

Cue Steve McQueen. California now has some 800,000 registered motorcycles, minibikes, dune buggies, gyroplanes—most of them within striking distance of the desert. Steve and his Wild Bunch roar out on weekends like those old Tetley Tea ladies who used to roar down the television roadways headed for the diner and their Tetley Tea fix. Fragile desert plants are uprooted; orange and blue markers are tied to plants to mark the way for trail rides. Months later the markers are still there, drooping in the hot air. Tire tracks are almost ineradicable in the dry desert air. General Patton, who used this portion of the desert for tank maneuvers during World War II, is forever memorialized: the scars of his tank tracks are still there, crisscrossing the silent sands like a multiplex and confused ticktacktoe game. The winner is a BMW R60. When a machine breaks down, it is junked on the spot. Why haul it back to civilization? In the desert it becomes target practice for city sharpshooters. When sharpshooters run out of targets, they then aim for the desert animals. Campers, waddling out by the thousands, pregnant with the fruits of consumerism's labors, drop a solid waste stream of junk and debris in their wake. (Not all transactions are deposits, though: a plank road, built by early pioneers across a portion of the desert and remaining for decades, has been almost destroyed by nature-lovers who ripped it up and burned it for firewood. Precious petroglyphs—ancient Indian rock paintings from the Stone Age—have, for the most part, been obliterated as the Machine Age freed man from his sprawling suburban shackles.)

As McQueen, who owned six motorcycles at last count, figures: "You end up pushing further and further back into the boonies trying to escape from other people and their noise and their crap," he says of his mechanized pursuit of the wilderness. "It's the problem that confronts all of us in a jam-packed world. Who are we running away from. The answer is simple: us." As for his desert playground,

McQueen knows the score. "I first began to understand it as a living thing back in my wilder days," he says. "I was interested in the Indians, and they had given me some peyote. I took off into the desert on my bike, bound and determined to whip it. I ran flat out, straight into the desert —I was all ego, challenging every bump and every gulch. The cactus ripped me up, the rocks chewed on my hide. I had sand in my nose and kangaroo rats in my ears. I rode until the bike ran out of gas and after that I just lay there. It was dead quiet, night falling and my bike making those little crackling noises as the metal cooled and settled. I knew then that not only could I never whip the desert, but that the whole thought of trying to whip it was the most ridiculous idea in the world."

When McQueen isn't trying not to whip the desert, he makes movies. And when he doesn't make movies he goes out to check on his businesses—one of them a plastics company.

Just in case a kid might grow up without knowing the joys of motorcycling, the Floyd Clymer Motorcycle Division has come up with a motorcycle for kiddies. "If your son can't pedal a bicycle, put him on an Indian mini mini Bambino," the ad reads. "This is an 18″ high motorcycle for the four to eight year old. The 50cc automatic clutch engine is de-tuned to hold the top speed down to 10–12 mph." Whatever happened to tricycles?

Nor are motorcycles the only beasts in the garbage field. Not at all. Our technology has produced a whole category now referred to as "off-the-road vehicles," which includes not only motorcycles but dune buggies and snowmobiles. At last count there were some 1,600,000 snowmobiles registered in the United States and Canada. Some 200,000 belonging to those desiring snow mobility are registered in Michigan, over 30,000 in Maine. There are even thousands of snowmobile enthusiasts in Nevada. Out they go, nature-lovers all. Sitting in their open cockpits, breathing in exhaust fumes (the exhaust is in front of some snowmobiles) and thrilling to spine-tingling outdoor adventure (most of the seats are unpadded).

Lloyd Clark, a Maine hunting guide, notes in worried Yankee tones that "when you come on any species of wildlife with the snowmobile, you'll find that species on the move, with a wild look of terror in his eyes." Beaver-trapping in Maine is reaching epidemic proportions as avid trappers, on the go with eyes aglint, race from one beaver trap to another on their snowmobiles. According to a report on vanishing wildlife, bobcats are being annihilated in certain areas by gamy sports-lovers who run them down on snowmobiles. Until Minnesota passed a law forbidding the harassment of animals by vehicles like the snowmobile, sportsmen used to run down foxes. Some chase moose and deer across the frozen fields until they drop from exhaustion. Then just a simple bang! bang! you're dead. Or sometimes it's even more simple: they just drop dead from exhaustion. No longer a moose—just garbage.

The Nevada Conservation Forum, a meeting of conservationists from that region, met in Reno last year, and among the topics on their agenda was the snowmobile. Not only does it harass the animals, sometimes ending in their death, it is a hazard to trees. "The tops of pine trees, sticking up from the surface of the snow, have been broken off by the snowmobiles," notes Dr. Harold Kleiforth of the University of Nevada's Desert Research Institute. All over the country, farmers and ranchers are furious at snowmobiles breaking down their fences and zooming across their snowy fields. Forest Service personnel at the Reno forum also complain bitterly about the litter and sanitation problem caused by snowmobile buffs who casually toss off beer cans, sandwich bags, broken thermos bottles.

And if mini-motorcyclists can have their very own children choppers—what about the kid who's a snowmobile enthusiast? Would American technology let you down, snow-loving kiddies? Don't you believe it. There is a "minimobile" available "built especially for children" by Lori Engineering of Southington, Connecticut, that "can provide all the fun thrills of a big snowmobile." Yes, kids, you too can scare-drive a full-grown moose to exhaustion.

Bad news notwithstanding, there seems to be no letup in snowmobile enthusiasm. After all, it is big business: sales are $1.4 billion a year and rising.

Not that technology growth does not have its avid boosters. It does indeed. Take Ayn Rand, the original 1950s trip. Students sat around the dorm and learned how to inhale Chesterfields over Ayn Rand. She was the pleated skirt of our mind. The burning question of the 1950s was: Is John Galt dead? And here she is again. Coming back to haunt us. We who have traded in our circle pin for an ecology button.

Now Ayn worries a lot about things. When the hippies and heads and radicals and revolutionaries and freaks of the late 1960s started stirring things up, Ayn got concerned. She saw a whole generation of new vipers emerging on the New Frontier, turning up their noses at progress, technology, and science, the very things that, to Ayn Rand, make America great.

To Ayn Rand, the self-styled savior of American youth, the whole ecology movement is a Commie pinko plot to overthrow decent law-and-order democracies everywhere and impose a global dictatorship. To her, the hordes of Earth weaklings were not marching through the land of the free and the home of the brave to let freedom ring through clean air. They were "young people who did not take the trouble to wash their own bodies, and [who] went out to clean the sidewalks of New York." As for the ecology movement, "the immediate goal is obvious: the destruction of the remnants of capitalism in today's mixed economy and the establishment of a global dictatorship. This goal does not have to be inferred—many speeches and books on the subject state explicitly that the ecological crusade is a means to that end." And we all thought it was just a simple recycling center. In reality, according to the logic of Ayn Rand, it is the front office of a global dictatorship. Tell *that* to the scrap-paper man.

Ayn looks about her and thrills with wonder and joy at the technological advances around her. And well she might. There is electricity and there are labor-saving devices. There is fast transportation. There are factories and assembly lines and a feast of products to make our life both better and easier. Nobody disputes that. Relax. Nobody wants to unplug you. The unwashed Armies of the Night of Technology are not marching to your house to disarm you of your automatic washing machine. Nobody wants you to trade in your dune buggy for a horse and buggy.

What they ask is simply that technology be made to work for people instead of against them. That technology, rather than mess things up, make things better. Rather than new cars every single year, how about a better mass transit system? As Stanford University biologist (and honorary president of Zero Population Growth) Paul Ehrlich puts it, "We need more transportation, but fewer automobiles. We need more housing, but less suburban sprawl. The world may need more aluminum and steel for communications networks, bridges and railroads, but the United States certainly does not need more beer cans."

Not that Ayn doesn't have a lot of clout on her side. One such big fist is none other than Dr. Lee A. DuBridge, science adviser to President Nixon. DuBridge once said, "I strongly reject the idea that we have to destroy our technological civilization, deflate and decrease the standard of living, to improve the quality of life. There may be a few who would like to return to the days of the caveman, but most of us believe that we live healthier, more pleasant lives than they did 10,000 years ago or even 100 years ago."

Dr. DuBridge is absolutely right. Everyone knows the quality of life goes up with the air pollution index. Think how much better our lives are today and then make some eco-freaking speech about the good old days. Think about the Cuyahoga River at Cleveland catching fire in 1969; the numbers of lives that brightened in just one night was heart-warming indeed. Think about the chances twentieth-century civilization provides its citizens. Could a caveman build his home on radioactive wastes? Of course not. Could he catch hot fish from thermally polluted waters? No way. Of course our lives are healthier. As Dr. DuBridge says, count your blessings, folks.

"No responsible observer has seriously suggested that technology be abandoned or that mankind return to hunting and gathering," says Paul Ehrlich calmingly. "Those who have rushed to technology's defense should stop grappling with this straw man and deal with the real issues: the focusing of technology on genuine human needs, the minimization of its adverse side effects, and the recognition that some problems do not have purely technological solutions. Few would deny that the sooner we abandon the technologies of war and planned obsolescence the better."

One of the best statements of the technological morass in which we are wading came in a public service ad by the Atlantic Richfield Company. "Man has created a technology that distorts his humanity," read the copy block. "We can shape a new technology that moves us forward toward a future which honors the human dimension. Suddenly, we know man's technological genius has not been an unmixed blessing. Blinded by its power to create material abundance, all of us allowed it to become an end unto itself and, in the process, to ravage the environment. We must balance technological and human factors. Working together, we can build a humanized technology that is our servant, not our master." Atlantic Richfield, on the one hand supplying the perfect pollution machine with the raw material to clot the air, and on the other producing some cogent thinking on technology vs. man vs. the environment.

And speaking of Madison Avenue: if we have supertechnology we must obviously have ways of getting rid of the products of our labors. Who wants an entire payroll working full speed ahead with fringe benefits and paid vacations and nothing to do with the damn stuff? Aha. Madison Avenue. To create garbage you have to sell it. And Madison Avenue is up to the assignment.

Madison Avenue is not Gregory Peck in a gray-flannel suit, striding out of Grand Central Station with a rising crescendo of the Hollywood Strings attached. Madison Avenue is more like a Gelusil tablet. There are more resident acid heads on Madison Avenue than the Lower East Side could produce in a lifetime. Madison Avenue is the overkill of the supersell. Madison Avenue is where Procter & Gamble leads the world in money spent on advertising. P & G doles out a total of $50.8 million a year in television advertising alone to tell us about detergents.

Advertising Age estimated that in 1970 business spent $20.8 billion on advertising. The one hundred largest corporate advertisers (with sales of approximately $225 billion) accounted for over a quarter of that, shelling out $4.7 billion in advertising, e.g.:

Procter & Gamble Co.	$265,000,000
General Foods Corp.	170,000,000
General Motors Corp.	129,765,000
Colgate-Palmolive Co.	121,000,000
Ford Motor Co.	90,250,000

Bringing up the rear, Company number 100—Mars, Inc.—
spent $10,875,000 on advertising to push Mars Bars.

Pollution has even become chic on Madison Avenue.
According to *The Sciences* magazine, the major industrial
polluters spent $1 billion in 1969 advertising their efforts at
pollution control. That is hardly comforting when one
stops to consider that that figure is ten times 'more than *all*
U.S. companies spent for pollution control devices during
the same period.

There then grew up, in the early 1970s, a field that became popularly known as "eco porn," which means, quite
simply, ripping off the environmental issue. To hear the
likes of the oil industry and paper companies tell it—in
their lush, four-color ads in the major magazines—they
were entirely devoted to building bird sanctuaries and forest preserves.

Not that it is all darkness and gloom and antipollution
hype on Madison Avenue. Periodically a little light-hearted
fun and frivolity creeps into the scene. "It was one of the
most beautiful accounts in the history of advertising,"
wrote Jerry Della Femina in *From Those Wonderful Folks Who
Gave You Pearl Harbor*, his wooly account of his gray-flannel
career on Madison Avenue. "We were selling sewing machines to Indians who couldn't run them because they had
no place to plug them in. We ran ads, we ran commercials,
and we made a lot of bread. If you got the Indian to make
the down payment, you were breaking even. The rest of the
stuff was gravy."

Bread and gravy. Fantastic. See the Indian. See the sewing machine. Look, ma—no plugs. Unless the Indians were
avid readers of *House Beautiful*, they probably wouldn't
figure out that the sewing machine could be turned into an
absolutely *divine* planter with ivy dripping from its zigzag
attachment. Another piece of solid waste, bought on time.

In their own very diverse and divergent ways, Ayn & Paul & Barry & Atlantic & Richfield are all quite right. There is really nothing wrong with American technology —just in the way it has been handled. "The answer to all of these environmental and resources problems is that we simply use less goods and services," says Charles F. Luce, chairman of the Consolidated Edison Company of New York. "In other words, that we get off this growth kick our economy has been on throughout the history of our country."

This is the kind of thinking that sets corporate teeth on edge. Get off this growth kick is like asking Roy Rogers to get off his horse. As Dr. Dennis Gabor, winner of the 1971 Nobel Prize in Physics, put it, "The insane quantitative growth must stop; innovation must not stop—it must take an entirely new direction. Instead of working blindly toward things bigger, it must work toward improving the quality of life rather than increasing its quantity." Unfortunately, he pointed out, "all our drive and optimism are bound up with continuous growth. 'Growth addiction,' is the unwritten and unconfessed religion of our times." As Dr. Gabor grimly sees it, "We have now reached a stage in which innovation has become compulsive. A large vested interest has been created, even apart from the military-industrial complex, embodied in the avant-garde industries and research organizations which believes that it must 'innovate or die.' "

Interestingly enough, Dr. Gabor is a staff scientist for CBS Laboratories, the folks who encourage new television sets every year and television shows to watch on them (*Bridget Loves Bernie* and *Stand Up and Cheer*). There is no justice in the world at all, is there? Only very unpoetic ironies.

"We've got about twenty years in which to reorganize," figures Dr. John List, assistant professor of engineering at the California Institute of Technology's new Environmental Quality Laboratory. "It's just plain affluence. It's growth per capita consumption. The only way out of it is to curb the consumption per person. Not exactly a no-growth situation, but slow it down."

Indeed, the slow-growth proponents seem to have a point. American affluence has reached staggering proportions. Nearly all of us have indoor plumbing, electricity, television sets, and cars. Personal incomes have more than tripled from a total of $226.2 billion in 1950 to $743 billion annually at the beginning of the 1970s. We have more money; therefore we spend more money to buy more things. In 1960 75 percent of all American households owned an automobile and 16 percent owned two or more. Ten years later, 79.6 percent owned at least one car, while those owning two or more had jumped to 29 percent. In 1960 4.9 percent of households owned a dishwasher; in 1970 that figure had gone up to 13.7 percent. Air conditioners were owned by 12.8 percent of households in 1960; by 1970 some 20.5 percent were choosing air conditioners over fans and cold baths.

We spend nearly $10 billion a year on tobacco, $55 billion on clothes, over $15 billion on alcoholic beverages. These figures are nearly 50 percent higher than they were ten years ago, nearly double the 1950 figures.

Not that everybody is taken in by the fancy talk of Messrs. Luce and Gabor, et al. Not at all. "The only healthy state of a nation is perpetual growth," said John L. O'Sullivan, a very well known and respected newspaper editor —in 1845. His very thoughts and ideals, however, have been echoed down through history by the likes of Yale economist Dr. Henry Wallich, who ringingly dismissed the no-growth ideas as "absolute bunk." As far as he sees it, that would be asking America to produce less, while working fewer hours and getting stuck with more leisure time. The

Yale doctor asks, what would happen then? "What are people going to occupy their time with? Religious contemplation? Art? Beautiful thoughts?"

Dr. Lamont Cole, ecology professor at Cornell University, made the point that so many ecologists, concerned citizens, and politicians—just about everybody but the economists and industrialists—have been making recently. "One of our basic errors," says Dr. Cole, "is that we always equate growth with goodness. Everything has had to keep growing—the population, the cities, the industries." Like a matron on a fudge binge, America has been stuffing more and more into her foundation garment. Until, finally, the final bit of fudge was force-fed in, and the foundation began to come apart at the seams. "We have to stop growth somewhere," Dr. Cole figures.

The very foundation—the environment—upon which we have built our magnificent Great American Dream Machine is crumbling. Where else does everything come from? And where else does it go? This was all fine and dandy until recently, when we began inventing things at such a pace, consuming things at such a rate, and throwing away things in such a state that they either cannot break down (there is too much of them) or will not break down (they cannot because we have formed new matter that breaks out of the completely closed natural cycle: plastics, aluminum, radioactive waste).

"The human race may be in even more trouble than we think," laments Cornell's Dr. Cole. "Very possibly, man won't know he has passed the point of no return until it's too late."

As New York City environmental chief Jerome Kretchmer sees it, "Concern about the environment challenges this concept of growth because the quality of life is nowhere factored into the standards by which we measure progress in America. We measure America's growth by the size of the Gross National Product. But the GNP includes output that pollutes and output that attempts to correct pollution—the pulp mill that ruins a river and the plant that tries to clean it up. This is an irrational yardstick." According to Kretchmer, at least $15 billion of our annual GNP goes to combat property damage caused by air pollution alone. "Is this growth?" he asks. His own reply: "This country needs a national policy on growth which would provide answers for some basic questions: How much growth is necessary? How much can we stand? What kind of growth will give the government the power to ban or limit automobiles, for example? At what point do we cut priorities toward growth and begin work on improving what we already have?"

"We are now paying the price for our need to conquer nature," Kretchmer said in the summer of 1971 as the environmental battles heated up around him. "Growth in our society has been defined in terms of quantity—bigger cars, more highways, taller buildings, sprawling cities—to the detriment of the quality of life." As Kretchmer sees it, "The conquering of the frontier made us arrogant toward nature and its limits and growth has been distorted into our modern idea of accumulating goods—quantity over quality."

According to the Population Reference Bureau, the world has doubled itself some thirty-one times since Adam and Eve. If we double ourselves up just sixteen more times, the bureau figures there will be one square yard available to each of us. One square yard on which to eat, sleep, live, breathe, pollute, play softball, chase butterflies, bake brownies, grow daffodils, and feed our pussycats. There is one bright spot the bureau notes: "The human population will stop growing long before 16 more doublings or even 8 more doublings, not for sheer want of space but for want of food, water and other natural resources and possibly because of environmental pollution."

America's current population is over 205,000,000. Some estimates put our turn-of-the-century population at 321,000,000 if we continue growing at our present growth rate. Zero Population Growth—ZPG—is an organization that requests a mere replacement level, i.e., each couple would limit themselves to only two children, enough to replace themselves. ZPG figures that even at that rate our year 2000 population would be over 250,000,000. And that is if every single couple in this country immediately switched over to the ZPG philosophy and acted accordingly.

No matter which figure you choose, we are stuck with a lot of people. As our population grows, our cities are sprawling farther and farther out. Los Angeles and suburbs (population 6,962,000) alone gobbles up seventy square miles of open land every year, creating what some are now calling "slurb"—sleazy, sprawling suburban subdivisions. With more and more people, our remaining open spaces become more and more threatened. A riot in Yosemite National Park in 1970 nearly forced its closing as people jammed in by the thousands over Labor Day weekend. Our beaches, from the air, look like nesting places for the world's fly population. Already 70 percent of our people live on 2 percent of the land, but even that figure is misleading. It sounds as if 98 percent of our land has remained untouched. Not true. Walk anywhere—you'll find a beer can. Remember Mount McKinley. Nor does it dramatize urban density—a high-rising philosophy that packs people by the pounds into smaller and smaller areas.

People want. People want cars that pollute the air, and necessitate the using of green nature to build gray ribbons of roadways in, and then end up as junk. People need housing and schools and hospitals and libraries and fire departments. To build these takes money and natural resources. To maintain them takes money. Already they are over-crowded and mostly inadequate. "It is difficult to see how we will be able to keep up the standard—much less im-prove it—for 100 million new Americans in the next 25 years," say the ZPG people. "People pollute the environ-ment by their mere biological existence," says ZPG.

The population explosion. The baby boom is the shot heard round the world. The hand that rocks the cradle also rocks the boat. The National Wildlife Federation, in its EQ (Environmental Quality) Index for 1970, figures it this way: by year 2000 America will have run out of both uranium and oil. By year 2022 we will have exhausted our natural lead supplies. By year 2026 we will have run out of zinc. The American Paper Institute figures we might run out of timber for wood and paper by 1980. And water. Good heav-ens. Turn off your taps, folks.[2] Estimates are that we are even running out of usable water. It is being polluted at such a rate that even the ecological Pollyannas are not hopeful for its immediate recovery. According to George Borgstrom of the University of Michigan, North Ameri-cans are removing twice as much water from their ground-water reserves as they return to it. Our EQ is slipping. (For example, it takes 100,000 gallons of water to produce an automobile. A single corn plant consumes 50 gallons of water in a growing season. A leaky faucet with a one-sixteenth-inch drip drops out four gallons of water in a day.)

Even good, gray President Eisenhower recognized the population problem, when in 1968 he remarked that "once,

2. Modern toilets use too much water when flushed. To remedy that, Governor Nelson Rockefeller's New York State environmental agency recommends put-ting a brick in the toilet tank to retard the water flow. One wonders: does Happy have bricks in her tanks?

as President, I thought and said that birth control was not the business of our Federal Government. The facts changed my mind. I have come to believe that the population explosion is the world's most critical problem." During his Senate campaign the late Senator Robert Kennedy whimsically confessed he was entirely disqualified to address himself to the problem as outlined by Ike. The next year in Peru he lightheartedly challenged the Peruvians—who have both one of the world's highest birth rates and one of the highest infant mortality rates—to outbreed him. "Deadly dangerous," mused *The Washington Post* of Bobby's challenge.

When *Time* Associate Editor Ray Kennedy (no relation to those other Kennedys) was pictured in 1970 in that magazine's Publisher's Letter, irate letters poured in from around the world. Kennedy was pictured with his family, a fulsome and freckled lot then numbering eight children, with wife Patsy pregnant with the ninth. "If each of Ray Kennedy's nine children should choose to follow their parents' example and their children after them and so on, for a mere nine generations," calculated one *Time* reader, "the country could have 387,420,489 Kennedys—or [nearly twice] as many Americans as there are today. Wherein lies the rationalization between loving children and simultaneously making the world a less desirable place for them to live? By having as many children as they want, the Kennedys impinge upon the future rights of their own blood to enjoy the good life they now prodigally enjoy." The letters came pouring in. *Time*'s Letters Department gave up counting. So, apparently, did the Kennedys: their personal population explosion, at last report, had plateaued at nine.

One of the more outspoken critics of population growth is Charlton Ogburn, Jr., author and a trustee of the National Parks Association. "Newspapers regularly report the plight—and complaints—of parents of twelve on relief," he once said, "without any suggestion that society has rights in the matter. Public figures who have become known partly because of their concern with the nation's future, like columnist Jack Anderson and entertainer Dick Gregory, can have nine and seven children respectively and not feel that they owe the public an apology any more than John Wayne, who also has seven children. The Governor of New Jersey, Richard J. Hughes, had three children by one wife, acquired three more with a second wife, and by her had an additional three."

The old argument that the affluent—like Anderson and Wayne and Governor Hughes and the various Kennedys—can afford to take care of the broods they breed is the argument put forth in justifying their profligate prolific inclinations. Of course they can afford to *buy* the physical things necessary for their own version of the good life. The point is that their excess paternalism just contributes to the general problem: that the United States of America with 5 percent of the world's population uses 40 percent of its natural resources and returns 30 percent of its pollution in thanks. The richer a family—or a country—the more they use and therefore the more they pollute. "Nearly all the problems we are wrestling with today," said Mr. Ogburn, "are being rendered far more difficult of solution by the addition of nearly 5,500 lodgers to the national boarding-house every day."

A new American is born every eight and a half seconds. That's a lot of Pampers, cribs, blankets, creamed carrots, tricycles, hair bows, school books, bicycles, gymnasiums, dormitories, automobiles, suburbs, hospitals, coffins, and tombstones.

"People are only part of the problem, and only part of the solution," say the folks over at Zero Population Growth. "But they are an inextricable part of both. No problem of the physical or the social environment can be solved if we have runaway population growth. The quality of our life is at stake. With population control we have a chance of solving our other problems. Without it, we can only hope to die fighting in a losing cause."

Just to keep the rapidly increasing numbers of cars on the roads, we are ripping off Mother Nature and paving over a million acres of oxygen-producing trees every year. (Well as she herself warns, "It's not nice to fool Mother Nature.") "Once you understand the problem," says biologist–ecologist Barry Commoner, "you find it's worse than you ever expected."

This country's annual population growth rate of 1 percent doesn't seem like much on the surface. A "1" seems such an insignificant figure. If, however, that current annual growth of 1 percent had been operating over the past five thousand years, the world would have a population of 2.7 billion people for *each square foot of land.* Somehow that is not my idea of fun. It is nice to have neighbors to share a cup of coffee with and borrow that mythical cup of sugar from. It is a whole other thing to have 2.7 billion of them standing on your square foot jostling for space.

Businessmen, however, see population growth not in terms of too much stress and strain on an already overworked world. They put population right into the marketplace, coming up covered with customers. "It's not just the number of babies—it's the same babies getting the same stuff from fifteen different people," said Marshall B. Sidman, president of the Kiddie Products Co. of Quincy, Massachusetts. "They buy and they don't even consider whether the baby needs it. That's what's so beautiful about our business."

"Overpopulation is at the bottom of almost everything," wrote *New York Post* columnist Pete Hamill late in 1971. "If there weren't so many people, and if they didn't continue to breed at such an irrational rate, then the cities wouldn't be so filthy, the schools so battered, crime so malevolent, and drugs sought so easily as a refuge from urban pressures. The streets of our cities could be safe gardens in which reasonable humans could lead sensible, civilized lives."

It is so oversimple, but so right. Jammed cheek by jowl onto that pitiful little 2 percent of our land we fight to breathe, fight to move, and, finally, fight to live. Life becomes a High Noon battle, and the only winner is the dark night of despair. Nothing is finite, as we are sadly discovering. Nothing lasts forever, most especially our precious environment, that thin, fragile blanket of nature that surrounds us. And people-mongers—male chauvinists, Catholics, lonely women, the whole mixed bag of them— are among the worst offenders. Marriage and the birthing of babies are two of the easiest things our laws provide for. Divorce, contraception, and abortion are among the most difficult. It costs a couple of dollars and a few minutes to get a marriage license. It costs thousands of dollars and endless bureaucratic and legal delays to get a divorce in this country. Having a baby is often as close as the nearest orgasm. Getting a prescription for The Pill or an abortion is often as difficult as ending a war or defeating still another SST proposal. Marriage and procreation are the single most serious activities of human existence; we make it as easy as catching a double feature on Saturday afternoon. Our tax laws are stacked in favor of more children; ditto our welfare rules.

The whole population issue reheated in early 1972 with the release of a prestigious study prepared by thirty-three eminent scientists in Britain who warned that to avoid environmental catastophe in the twenty-first century, Britain must cut her population by half, stop building roads, and tax the use of power and raw materials. "We cannot think of it in linear fashion—as if the next 1,000 years would be like the last 1,000 years," said one spokesman of the *Blueprint for Survival.* Britain's current population is 55 million; the *Blueprint* favors an optimum population of 30 million and suggests aiming for that figure over the next 150 years. "If current trends are allowed to persist," the *Blueprint* states, "the breakdown of society and the irreversible disruption of the life-support systems on this planet—possibly by the end of the century, certainly within the lifetime of our children—are inevitable."

Those were strong, sober words from one of the leading scientific communities in the world.

Are we a world without solutions? In Britain, where abortion has been legal for some years, a series of television commercials on birth control were begun in 1971. "Economic levers are also available," says Charlton Ogburn. Instead of tax incentives for people who have children—$600 exemptions and all that—why not a tax break for people who abstain? A bonus for family banning. "Fines, proportionate to the offender's capacity to pay, could be levied against parents for each child they produce in excess of two; beyond a certain limit the offenders could be deprived of the right to vote." Well, it's a bit harsh, but as Ogburn angrily figures, "Why should those indifferent to society's future be given a voice in it?" Beyond that, states can repeal abortion laws and disseminate both birth control information and contraceptives.

As for the Vatican, old Mother Church clucking her tongue and clacking her beads over not only her own children but all God's children, Ogburn even holds hope there. "The Vatican has changed its mind in the past, and can and must change it again."

Indeed the whole idea of limiting people is catching on. The Pope might not be dispensing birth control pills along with his Papal Blessing at Saint Peter's, but the church fabric has some very severe rends because of the birth control issue. Depending on which priest a woman consults, she may get very liberal birth control information and priestly permission to go forth in all good conscience and have intercourse with her husband while downing an Enovid every morning. In May of 1971 twenty Protestant theologians met in Claremont, California, for a three-day symposium to "consider the religious dimensions of the ecological issue." Their verdict: "Theology, like Western philosophy, has gone too far in making Man the center of attention."

In 1971 the U.S. Census Bureau reported that for the first time in the history of the country the fertility rate had dropped to the bare replacement level—the threshold of Zero Population Growth—at 2.1 children per woman. More cautious voices, however, warned that the rate could rise just as quickly as it fell. "There is still a bomb there," said demographer Philip M. Hauser of the University of Chicago. "Basic changes in reproductive behavior must be measured in generations, not years." From a postwar high of nearly four babies per family (reached in 1957, which made it the highest birth year since 1917) the trend downward was not such a plummet as some would have it and the slight drop is attributed in part to more widespread contraception, the further spread of legalized abortion, and the rising numbers of single women. "It would be nice for me to be able to say that our propaganda is paying off," said ZPG's Ehrlich, "but that would be reckless. These reports are extremely encouraging. But what has come down quickly, can go up quickly." Just like miniskirts and midiskirts.

And so the battle rages. And while the U.S. baby rate dropped from four to three to two, the world population continued to rise and, according to the Census Bureau, will do so for at least the next sixty years.

On the other side of the population problem, Barry Commoner has calmly and lucidly analyzed pollution and figures that neither affluence nor population has much to do with it. He points a long and lingering finger at technology. But without both the money to buy the stuff (affluence) and the market to peddle it to (population) where would technology be in the world of pollution? Even Commoner sense should figure out that New York's streets and sidewalks, for example, would be cleaner if there were fewer people littering. That fewer pop-top cub-size aluminum cans would be sold if (1) so many people did not have the (2) money to buy them. Technology—at least American technology—produces only what the market will bear. And if the market is big (people) and bullish (affluent), so is production. And, unfortunately, so is pollution.

We have lived long with the idea of population proliferation. Machismo perpetrated Momism until we are where we are today: teetering on the point of no return, behind us, the affluent generation snapping at our heels; ahead of us, monumental decisions to be made. About population, about technology, about affluence. About our environment. Which is to say, our whole way of life.

Perhaps the most controversial report to be issued recently was a study done at the Massachusetts Institute of Technology under the auspices of the Club of Rome, a group of scientists and intellectuals who banded together in 1968 to study ways of averting what they felt was the impending breakdown of society caused by uncontrolled growth and overpopulation. According to the 1972 study, *The Limits to Growth*, if the world keeps going and growing at its present breakneck speed, by year 2100 food and industrial output will drop to almost nothing while our natural resources will have been nearly exhausted. Conversely, with some sort of regulation, population could be stabilized along with industrial output, which would end pollution and save the food supply and our dwindling stock of natural resources.

"Our view is we don't have any alternatives," said Dr. Dennis Meadows, who directed the MIT study. "It's not as though we can choose to keep growing or not. We are certainly going to stop growing. The question is, do we do it in a way that is most consistent with our goals or do we just let nature take its course."

According to Dr. Meadows, if we simply let "nature take its course" we are doomed. Population would drop owing to starvation and disease. It is a scary scenario:

—With growing population industrial capacity rises, along with its demand for oil, metals, and other resources.

—As wells and mines are exhausted, prices go up, leaving less money for reinvestment in future growth.

—Finally, when investment falls below depreciation of manufacturing facilities, the industrial base collapses, along with services and agriculture.

—Population plunges from lack of food and medical services.

The MIT team studied various alternatives and came up with a system they believe capable of satisfying the basic material requirements of mankind without causing a collapse. To them, such a world would require:

—Stabilization of population and industrial capacity.

—Sharp reduction in pollution and in resources consumption per unit of industrial output.

—Introduction of efficient technological methods: recycling of resources, pollution control, restoration of eroded land, and prolonged use of capital.

—A shift in emphasis away from factory-produced goods toward food and nonmaterial services such as education and health.

"A society released from struggling with the many problems caused by growth may have more energy and ingenuity available for solving other problems," says Dr. Meadows.

We are indeed facing a revolution in this country, involving technology, our entire economic structure, and every priority we hold near and dear. It is a revolution that will —as it is already beginning to do—affect every one of us. Right now it is kicking away at the very foundation upon which we have built what was once considered the soundest system in the history of the world: free enterprise. For if we are to win this battle and close this garbage gap we are busily building, we must, out of necessity, restructure everything. We must reorder our priorities. We must turn our technology around and make it work for us, not against us. For if we can get a man to the moon, it is not so unreasonable—nor so corny as some may think—both to ask why we can't get our garbage efficiently picked up and put down, and to *demand* why. Demand clean air and clear water and a quiet atmosphere in which to listen to our own thoughts. We must question our affluence that allows us to buy frivolous goodies wrapped in expensive and excess packaging, while citizens of this, the most affluent country on the face of the earth, can live and die underclothed. For what earthly good does it do us to send men to the moon and destroy villages in Southeast Asia with the products of our technology if, at the same time, our own backyards are cluttered with our neglected priorities.

What good does it do to go on producing the means to the good life—cars and planes and Christmas presents and office towers and electric curlers and throw-away cigarette lighters—when the end result is an environment so blighted we cannot enjoy the good life we have worked so hard to build.

If things come to such a pass, perhaps there will indeed come a time when our affluence is severely limited. Income taxes will be aimed at wiping out a standard of living that in itself threatens to destroy us. Tax breaks will be given for the environment, and not against it. It will be considerably cheaper to recycle paper, instead of chopping down trees. We will return bottles instead of throwing them away and making new ones. It will be made cheaper to clean up at the source of pollution rather than after something has already been dumped, discarded, burned, or buried. Incentives will be given for fewer children instead of more. There will be bounties and taxes on excess packaging. Luxury taxes will mount up to discourage the purchase of such items, making capital available for necessities. For if the apathy that malingers in some sections of this country—most notably in the business sector that is always and piously "responsible to our stockholders"—does not disappear of its own accord, there can be no recourse but to legislate good taste for the good life.

The problem is too complex to narrow the causes down to only technology or affluence or population. The problem is a frightening multiple-choice question; and the answer is all of these, not merely one of these.

IV

Aunt Jemima Is Unfair to the Environment

Although not the biggest culprit,[1] packaging is emerging as the arch villain in our solid wasteland. Packaging is now about 14.5 percent of that 360 million tons of garbage we create each and every year. Even pre-modern math tells us that that percentage of 360 million tons comes to 51.7 million tons of throwaway packaging per year.

The effluence of our affluence is a plastic aspirin bottle stuffed with cotton, resting inside its very own cardboard box; vegetables tucked cozily in plastic, sitting on individual cardboard (or, sometimes, plastic) trays. We have self-

---

1. Paper, in all forms, constitutes 50 to 60 percent of all municipal solid waste.

baking aluminum trays for our corn muffins. Fortunately, we eat the muffins and throw away the aluminum pan. We have plastic cook-in vegetable bags. We have individual plastic pudding cups and individual margarine tubs. Throw away the bags, the cups, the tubs.

If we are dedicated to garbage in general, we must surely note our specific dedication to packaging, for if there is one thing we love, it is our one-way, throwaway, disposable package. One would think we consumed just for the sheer animal pleasure of ripping off all that packaging material and throwing it away. In 1958 every American used (which is different from "used up") 404 pounds of packaging materials. A modest effort indeed when you consider that by 1966 our consumption was up to 525 pounds of packaging per person.[2] Indeed by 1976 we are expected to contribute 661 pounds of throwaway packaging per year. In less than twenty years—from 1958 to 1976—we can expect to increase our packaging proliferation by more than half.

And our packaging problem is getting bigger. The question then becomes: can we continue to live in a neighborhood where the bad guy weighs 51.7 million tons and is still growing? For it costs a fortune to keep packaging around. In 1939 the value of packaging shipped out was $2 billion. By 1968 that value had increased to $18 billion; three years later in 1971 it was nearly doubled—approximately $31 billion. That can of beer you bought? Of its total cost, at least 43 percent of it is for the packaging. For the flip-top and crushable can. Perhaps it is time to rethink our beer-drinking philosophy. To chug-a-lug beer in a cub-size can costs

2. For the record, our packaging consumption works out this way: of the 525 pounds of packaging we use per year, we go through 255 pounds of paper (remember—this is packaging paper. Not the St. Louis *Post-Dispatch* or *Screw* or copies of *My Weekly Reader*), 72.6 pounds of metal, 83.6 pounds of glass, 11.2 pounds of plastic, 41.2 pounds of wood, and 2.6 pounds of textiles. Plus various and sundry miscellaneous mixtures that total 58.3 pounds.

more than doing it the old-fashioned ten-ounce way. A six-pack of nonreturnable cans of beer runs about $1.25 instead of 95 cents for a six-pack of returnable bottles. As for our soda pop, the New York City Environmental Protection Administration says it is 95 percent water anyway, so all we are paying for is 5 percent flavoring and fizz, and a container. Nobody is safe: 36 percent of the cost of baby food is for the glass jar. And who do you think pays for those little plastic tubs around individual portions of margarine? Who do you think pays for those little pudding cups and plastic wraps and aluminum trays? Not Aunt Jemima or Mr. Fleischmann or Sara Lee.

That's only as it should be, according to Arsen J. Darnay of the Midwest Research Institute. After all, he reasons, it's the consumer who buys it and the consumer who throws it away. Why shouldn't he pay for it? "It is my personal view that the difficulties packages are said to cause are not the responsibility of those who produce materials," said Mr. Darnay at a packaging meeting in 1969. Environmentalists quickly learned that he—a devil's advocate of the first magnitude—was most assuredly not on the side of the angels. As far as Darnay was concerned, the cost of picking up packaging waste should be borne by the generator of that waste, and "in the content of packaging wastes, the generator is the *consumer*, not the *producer.*" Since nothing is borne free, the consumer once again found the market shelves stocked against him.

"Modern industrial technology has encased economic goods of no significantly increased human value in increasingly larger amounts of environmentally harmful wrappings," says Dr. Barry Commoner. "Result: the mounting heaps of rubbish that symbolize the advent of the technological age." Or, as New York City's Jerome Kretchmer puts it very succinctly: "Excess packaging adds to our tremendous solid waste problems and serves no useful purpose other than to create profits for some industries." To that end, Kretchmer figures "we must go to packagers and make them think about where their products end up—and at what cost."

Of all the restrictive legislation, however, the biggest and best is in Oregon. There legislators passed a bill—the first statewide bill in the country—requiring a nickel deposit on all returnables, plus a two-cent tax on no-returns and an outright ban on pull-tabs. In California, South San Francisco banned no-returns in July of 1971. Others on the no-return ban-wagon at the end of 1971 were Howard County, Maryland; Princeton, Deephaven, and East Bethel, Minnesota; Edgewater, N.J.; Sterling Heights, Michigan, and Loudoun County, Virginia. No-return bottles were also banned in Wayne City, Livonia, and Troy, Michigan, as in Barberton, Ohio, and Northfield, Vermont.

(Actually, Oregon's statewide taxation of nonreturnables is just part of the parcel of environmentally sound legislation passed recently in that state. Additionally, they passed the first property tax advantage for land devoted to outdoor recreation rather than real estate subdivision, set rigid rules governing the siting of nuclear power plants, removed more than three thousand billboards across the state, and set aside $1.3 million in state highway funds to construct hiking and bicycling trails throughout the state. And besides that, Oregon Governor Tom McCall encourages people just to visit Oregon. "But for heaven's sake—don't move here." He figures they have enough people without importing them.)

Not that everybody is sympathetic to the horrors of excess packaging. In a talk given at a White House Seminar on the Environment, Judd Alexander, vice-president for environmental affairs at the American Can Company, made this point: "Now, when housewives show up in droves at state legislative hearings to testify against packag-

ing, I sometimes wonder if we should have been so successful. If we had not been able to reduce food preparation time
by two-thirds those lovely ladies would not have *time* to
offer testimony against packaging. They would be home
shelling peas or husking corn."

Most ecologists and concerned consumers agree that
plastic is the biggest troublemaker of them all. Plastic, in-
sist the environmentalists, does not burn up efficiently in
this country's pathetic incinerators. Instead, plastic melts
into glutinous globs, gumming up the already inefficient
incinerators. Secondly, it is not biodegradable. It does not
break down naturally in dumps and landfill (like an orange
peel or an eggshell or a love letter). Plastic just sits there,
staunch and sturdy, a resinous lump on the landscape. Who
would have thought it of that harmless-looking Baggie?
That disposable diaper, that styrofoam coffee cup? Plastics[3]
are indeed an irritant on the unprotected flank of the environmental body.

Not that the plastics industry is not up to all the flak is
now being hurled at them by the environmental guns.
When the incoming rounds of bad press began to get too
heavy, the industry organization—the Society of the Plastics Industry, Inc.—began to fight back. "Sure, we're a
polluter," Ralph Harding, executive vice-president of the
society, admitted. "So is everybody else. We've been getting far more attention than we think we deserve. Why, the
PVC bottles (polyvinyl chloride)[4] have fallen into the role
of sacrificial lambs, dramatic examples used to focus attention on solid waste problems."

There stand the plastics of our solid wasteland—poly-

---

3. According to the Bureau of Solid Waste Management of HEW, some 4 billion
pounds of plastic packaging were used in 1970. By 1976 it is estimated to rise to
over 6 billion pounds.
4. PVC bottles generally hold such items as shampoo and medicine.

ethylenes and polyvinyl chlorides and polystyrenes and polyurethanes—all 4 million pounds of it. As one government report put it, "Plastics are where it's at."

Plastic is good, the SPI insists, the best thing yet for sanitary landfill because—turning the environmentalists' favorite argument around and using it against them—it is not biodegradable. The very fact that it just sat in the landscape thumbing its plastic nose at us made it, in the eyes of the plastic industry, the best of all possible worlds.

All around the country, plastic industry spokesmen were insisting plastics *burn better* than anything else because they fire up fast and set everything ablaze more efficiently. And so what if PVC (polyvinyl chloride) gives off hydrochloric acid? After all, "hydrochloric acid is the same acid contained in the human stomach."

Just to add credibility and credence to its arguments, the SPI likes to haul out the results of a research project conducted by two chemical engineers from the School of Engineering and Science at New York University. It is a heady little report entitled *Municipal Incineration of Refuse with 2% and 4% additions of Four Plastics: Polyethylene, Polystyrene, Polyurethane, Polyvinyl Chloride.* The report made plastic sound like Santa Claus. According to the report, not only does plastic not gook and gum up incinerators, but incinerator operators were jubilant whenever they got a load of plastics for their front burners. Plastics burn with such a high heat that they tend to burn up everything near them —wet garbage, damp paper, soggy foodstuffs. And as for air pollution and soot—not a trace of it.

It sounded great. The two NYU scientists used the municipal incinerator at Babylon, Long Island, plus the four main types of plastics most likely to appear in a city incinerator: polyethylene (used to make most plastic containers for detergents, bleach, etc.), polyvinyl chloride (most heavier, clear plastic bottles such as shampoo bottles and medicine bottles), polystyrene (meat trays, egg cartons, styrofoam cups, etc.), and polyurethane (cigarette filters, etc.).

The engineers loaded the incinerators with garbage containing four times the average level of plastics and fired it up. Things went beyond the wildest dreams of the plastics industry. The plastic set everything else in sight on fire. According to their published report there was no air pollution and no dangerous halogens released to poison our atmosphere. No wonder: SPI funded the project, hired the chemists, paid them.

But, as we all know, where there's black there's white. Where there's one side there's the other. There is independent research on the plastics problem now facing this country's incinerators and landfills. One of the best is *Environmental Effects of the Incineration of Plastics*, done by R. Heimburg of the Syracuse University Research Corporation. Dr. Heimburg's report is short and to the point: "Plastics [polyethylene, polystyrene] burned with cleaner, less hazardous effluents than most 'natural' products such as coffee, sawdust, or rice." But—and there is always a *but*—only when burned under absolutely optimum conditions, i.e., topnotch incinerator with topnotch performance. Translation: brand-new. With fully 75 percent of this country's incinerators labeled substandard by the U.S. government, that is hardly encouraging.

"Our present research has indicated that some plastics are innocuous, some just bothersome, some very dangerous to health when incinerated." But Heimburg stressed, conditions in all cases have to be absolutely optimum. And just in case nobody's worried, "PVC is dangerous even when burned under optimum conditions."

PVC—shampoo bottles, medicine bottles, and the like—strikes terror and horror in the hearts of all environmentalists out lobbying in the landfill for the greening of America. "PVC" is to an environmentalist what "f*ck" is to the local librarian: something to be banned.

While optimum burning of other plastics may yield few environmental problems, rest assured PVC will make up for them. "Young plants died within one minute from the beginning of the exposure," and rats had very gross reactions indeed, with "heavy breathing and burning of the eyes." In another test done by Heimburg, "(A) . . . The eyes [of the rats] were swollen almost shut, accompanied by bloody discharge from that area. (B) There was sometimes a white film covering the cornea. (C) The nose was reddened and often a bloody discharge occurred there also. (D) Usually a gray, mucous-like discharge, eventually covering much of the rat's ventral (abdominal) surface." Autopsies showed "enlargement of the lungs with hemorrhaging" present.

Heimburg's conclusions indicated that "combusting polyethylene, polypropylene or polystyrene seems to pose little or no threat to the environment, except possibly via fillers or soot" but "polyvinyl chloride yields hazardous effluents no matter how incinerated."

Well, so much for medicine bottles and shampoo bottles.

Again, the point is not packaging but *excess*, frivolous, gimmick packaging. Who needs plastic egg cartons when the good old-fashioned cardboard ones (made of recycled paper) work just as well? Who needs plastic wrap around cucumbers and turnips and ears of corn and cantaloupes and carrots and eggplants and apples and potatoes when they can be sold loose? These things are not so fragile that a housewife bearing down on them—shopping list raised on high—is a menace when she picks them up and feels them.

And who needs all that plastic and cardboard around a single plastic bottle of vitamins? The consumer just (1) pays for the packaging itself, then has to (2) haul it home and (3) tear it off and (4) load up his own garbage can with it. From there, it is hauled away to (5) become part of that $3.7 billion disposal bill we all have to foot every year.

Who needs those cook-in pouches? Who needs pieces of plastic in between his cheese slices? Who needs plastic "windows" in cardboard boxes? If we don't know what macaroni looks like by this time, it is time to lock ourselves out of the supermarkets voluntarily. Who needs plastic berry trays and tomato trays? Why not thick old cardboard, just like the ones Mom's berries came in? Nostalgia, in some forms at least, is good for us.

Once the plastics industry was attacked it fought back with vigor indeed. "We don't have to be on the defensive all the time," said Harding, of the Society of the Plastics Industry. But they were. "What about all of the other packaging materials," Harding asked, "almost all of which are 'one-way'? Instead of solutions, some people prefer to find scapegoats. Today the plastics industry is unfortunately emerging as one of the favorite targets for misinformation and uninformed criticism."

Some environmentalists prefer to lash out at metals, which are usually thought of in terms of beer cans. All over the country cans are stacking up around us, but they are not all beer cans. Think of the family cat and the family dog. Think of creamed corn and canned peaches. Think of underarm deodorant and aerosol room freshener and bug spray. Think of floor wax and tomato juice and jellied cranberries. Just look at all those metal containers keeping those beer cans company. And that doesn't even include aluminum foil (think of the Thanksgiving turkey) and aluminum tins (think of Sara Lee banana cake). The whole pollution problem is not sitting in our rumpus room ripping off his fair share of the market's pop-tops. Indeed not; the pollution problem is also sitting in our kitchen, in our bathroom, in our bedroom. Indeed, all over our mortgaged split-level.

Yes, that feast of garbage to which we have so willfully and willy-nilly committed ourselves contains a sizable amount of discarded metals. We Americans purchase and consume the contents of some 62 billion cans a year. In a year, the average American empties about 302 cans (about 6 a week). That prorates out rather nicely in that we can empty our allotted can on a one-a-day basis Monday through Saturday, resting up from our labors on Sunday.

(Interestingly enough, the common tin can, which we see resting and rusting so poetically along our roadsides, is not really tin at all. For the most part, the tin can is a steel can. Political upheavals in Bolivia and Malaysia, the major sources of tin, caused the prices to fluctuate so wildly that American technology applied itself to the problem of the tin can and came up with a solution that is not dependent upon the politics of Bolivia: the tin-free steel can.)

Because of the high visibility of cans—usually called "litter"—the can industry has been sharing its load of verbal garbage with the plastics industry. The public outcry over nonreturnable containers has been swift and loud.

"Any one of us could help reduce pollution if we would just drop dead," is the succinct answer of Judd Alexander, vice-president for environmental affairs at the American Can Company. But, wait—he is only joshing. "Somehow," he admits, "I just cannot bring myself to make the sacrifice." Good of you.

Cans are a somewhat different problem to environmentalists than, say, plastics and paper in packaging. A can is usually simple, functional, and very necessary, and without the frills and fuss of plastic packaging. Even eco-freaks will admit that a can cannot be classified as a frill. Your peas and corn and sauerkraut and pumpkin-pie filler and cat food and dog food have to come packaged some way.

Some people suggest we opt for fresh fruits and vegetables. And instead of feeding the family pet canned food, why not switch to fresh food? Besides, according to Dr. Barry Commoner, most canned cat food contains protein-rich anchovies that we literally steal from the Peruvians by fishing off their coastal waters. It is Dr. Commoner's contention that the undernourished Peruvian peasants could better use those anchovies—brain food—than our family cats. "And we don't even eat the cats!" he says. Much canned dog food is made from the fast-dwindling herds of wild mustangs who roam our western ranges.

The Pop artists among us might find great cultural and aesthetic joys in simply looking at a well-designed can of Alpo or Ken-L Ration. The sight of a colorful can of horse-meat and meat by-products for Rover or Spot or Lad is sheer heaven to them: graphics, four-color labels, an all-steel can. Others, however, find more joy in the sight—or the thought—of a wild mustang galloping free and unfettered.

The thing about the tin men of our society is that they are no less environmentally offensive than the plastic man, but they are more subtle in their response to the attacks that are piling up around them. They approach the problem in that good old American way: a little hype goes a long way. In 1971 both the American Can and Continental Can companies admitted to spending $1 million each on the "Can People" ad campaign that posed the question: "Cans. Bad Guys or Good Guys?" and, of course, answered the question that they are most assuredly the Good Guys. One million dollars each on a campaign that the Council on Economic Priorities, an independent watchdog group, said was designed to "divert attention from the actual impact of the staggering volume of container wastes." The Can People ads concentrated on litter and tsk-tsked the American public for confusing their cans with litter. They pointed out that "of all the litter on roads and highway, 83 percent isn't cans." But: 17 percent of it is.

(When the Environmental Action Coalition asked Continental Can for some money to finance EAC environmental projects, Continental Can dipped into its pocket and came up with $3,000 plus hundreds of bags stamped "The Can People." It was a cheap way out, compared to that $1 million Can People campaign.)

Despite the disclaimer by the industry, litter is a mammoth problem. When the State of Florida did a concentrated roadside cleanup, it discovered to its horror 8.1 million beverage cans loitering on the state's roadsides. That calculated out to some 8,000 beer and soda cans per mile in Florida, a lot of litter.

Beyond the aesthetic cost of containers, there is also a staggering pick-up and delivery cost once those containers become garbage. In 1969 an estimated 78 billion cans and bottles entered the U.S. solid waste stream. Seventy-eight billion. By 1980 that figure is expected to rise to 100 billion. By the mid-1960s we were already paying out $122 million annually to pick up and deliver these containers to the dump. Yonkers, New York, for example, sat down and figured out that its 204,000 residents paid over $1.2 million a year to dispose of the 5,200 tons of cans and bottles that accumulated in the city.

Not that those nonreturnable, crushable, throwaway aluminum cans so dear to the hearts of the superconsumer are not without drawbacks. Those flip-tops *hurt* when you step on them. So do the stitches. And those individual pudding cups? Consumers Union—the watchdog group that publishes *Consumer Reports*—discovered the flip-tops were really the rip-tops. They cut. The cans, big favorites of the school lunch brigade, were designed so that come lunchtime the schoolchild zips off the lid by its ring-pull. Then what? Well, pudding sticks to the can lid, and any

American school kid worth his ABCs does what comes naturally: he licks the lid. And cuts his tongue. Or cuts his fingers on the inside of the rim. "We found we could slice a deep gash in a raw chicken leg with the lid," CU reported.

Hunt Snack Pack, first to jump into the marketplace armed with individual pudding cups and one of the top money-earners in the field, began getting complaints. Edward Gelsthorpe, then president of Hunt-Wesson, did what any corporation would do: he launched a safety campaign. He did not remove the cans from the shelves until they could be made safer—he launched a safety campaign, with magazine ads and classroom posters showing the proper and safe way to open the ring-pull can. Some mothers figured they would rather have their kids learn to read something other than Hunt-Wesson safety posters. The cardboard carrier for the multipacks of pudding cups also carries a warning of the dangers, and a pitch to "Keep America Beautiful" by not littering with the can. Additionally, Mr. Gelsthorpe offered a free plastic spoon to keep the kiddies from lid licking.

Besides the dangers of cutting, this new entry into the highly competitive convenience-food field costs fifteen to seventeen cents for a single serving, as compared to about eight or nine cents for packaged pudding mix. "Is the convenience worth the price?" CU asks. "Is the convenience of the cans worth their hazard? We think not," the group concluded.

One of the most obvious targets of ecologists is what they consider to be the excessive use of aluminum in packaging. Individual pudding cups, cub-size beer cans, self-bake aluminum pans. Aunt Jemima became a particular target for one New York-based environment group. A member of Consumers Lobby for the Environment, pushing her shopping cart through her local supermarket aisles, was appalled when she saw a package of Aunt Jemima corn muffin mix, complete with its own aluminum baking tin: bake in it, throw it away. The whole naughty cycle spun threateningly in front of the shopper's eyes. Aluminum ripped out of Mother Nature, made into a self-bake aluminum pan for which the customer paid extra in hidden costs, and which she had to carry home with her and then discard, thus contributing to her own garbage load plus the disposal problem: burn it or bury it.

The lobby member sat down and wrote a mildly outraged letter to Quaker Oats, Aunt Jemima's parent company. "Surely most kitchens these days come equipped with their own baking utensils," she wrote, "thus making it unnecessary for you to provide individual aluminum baking tins." Besides, "aluminum cannot be reduced by incineration nor will it break down naturally in sanitary landfill, the two major methods of solid waste disposal now being used in this country. Since we are fast running out of landfill in our metropolitan areas, we would like to protest such unnecessary and irresponsible use of aluminum. In addition, you are wasting precious natural resources by this wasteful use of aluminum."

The windup: "Aunt Jemima is unfair to the environment."

It sounded like a watchword to be carried to every nook and cranny of every supermarket: "Aunt Jemima is unfair to the environment."

Would Aunt Jemima let us down? "The preservation of our environment is important to all of us," wrote the company, "and we are pleased to have this opportunity to discuss it with you." As far as they saw it, they were in the clear. "We don't feel that the aluminum pan is an insurmountable problem. Aluminum can and should be recycled." The point being it was up to the consumer, *not* Aunt Jemima.

In the good old days a few years back, we all lived in a Coca-Cola billboard. The Girl Next Door with her bouncy blond pageboy and her blushing Avon cheeks and her Fire and Ice nail polish and her pleated skirt and saddle shoes and bobby sox was drinking Coca-Cola out of a returnable glass bottle. Ah, yes, things were more familiar then. You knew the bottle was glass and you knew it was returnable. You even knew it was 6½ fluid ounces, because that's all we had to choose from. But today *nothing* is that predictable. The Girl Next Door moved out of the neighborhood, and who knows what she is wearing. As for those Coca-Cola bottles—chances are they are aluminum. Or, at best, no deposit–no return bottles ranging anywhere from 8 ounces (if you can find those antiques anymore) to a full gallon.

Ah, yes. Times are a-changin'. Gone are the good old days. The glass industry is finding itself adrift in a shattered marketplace. Plastic is coming up fast because it is lightweight, indestructible, and—best of all possible worlds—plastic prices are becoming more and more competitive. Aluminum cans are taking a lot out of the beverage market, which once used to be entirely dominated by glass containers in those familiar old returnable bottles. Still, glass accounts for 6 percent of the garbage we produce each year. And 6 percent of 360 million tons is an awful lot of glass. (Translation: 21.6 million tons.)

Now, admittedly glass has a lot of problems to contend with. To make a returnable container strong enough to withstand both the Teamsters Union and vending machines and the pitch of excitement at, say, a baseball doubleheader, a glass bottle has to be strong. That means it weighs more and uses more glass and is more expensive. Until recently the glass industry did just that: made a heavy, durable bottle that made an average of forty round trips per single bottle. Then, with aluminum and plastics invading the marketplace along with the whole rush to consumer convenience, the bottle industry geared itself up to dispose of the returnable bottle. To manufacture a container to compete with throwaway aluminum and plastic, the glass industry discovered it could produce a lighter-weight bottle, then give it a protective coating to offset any destructive tendencies that came about as a result of the lighter-weight bottle. *Voilà!* The no deposit–no return bottle. More solid waste.

The role of the one-way glass bottle as an ecological villain is easy enough to see: if you don't return it, you throw it away. If you throw it away you (1) add to the already enormous garbage heap; (2) destroy more natural resources needed to replace the discarded bottles with new ones. Although the glass container market has suffered seriously at the hands of aluminum and plastics—it seems to have stabilized and leveled off rather than doing the all-American thing of growing—it is still a viable industry. Some 43 billion glass containers are used each year. Most of them one-way, no deposit–no return, throwaway glass containers.

A government report, *The Role of Packaging in Solid Waste Management 1966 to 1976*, summed it up in depressingly dry terms: "Three factors are bringing about the switch to nonreturnable containers: 1) the consumer's preference for a container which need not be returned to the retail establishment; 2) the retailer's disinclination to handle returnable bottles; and 3) the packaging material manufacturer's desire to exploit the potential of the beverage container market to the fullest."

What that simply means is that we prefer to rip off our natural resources to produce throwaway bottles that in turn rip off our natural resources by turning them into garbage dumps. All because the manufacturers are pandering to the throwaway mentality they have encouraged.

Government researchers predict that from 1966 to 1976 non-returnables will increase from 25.6 billion containers to 58.1 billion units, an increase of 127 percent. Returnables will drop from 2.7 billion units to 1.7 billion units during that same ten-year period.

Meanwhile, the campaign to promote disposables at the expense of both returnables and the environment escalated to dizzying heights in 1971 and 1972 as environmentalists and legislators coalesced around the disposable issue. In Washington State, for example, an intensive industry campaign was waged to defeat "Initiative 256," which would have required a nickel deposit on soda-pop and beer containers. The referendum garnered more than twice the necessary signatures to put it on the ballot, while polls showed that 80 percent of the voters favored just such an environmentally sound levy. Then industry put on its combat boots, marched in, and mounted an expensive ad campaign that —naturally—soundly defeated the initiative. Money talks, and a hype-sensitive citizenry listens. William Rodgers, a law professor at the University of Washington, figured that the group supporting "256" had about $6,000 at their disposal to advertise their goals, while the industry-sponsored groups had some $600,000 with which to fight the nickel deposit. It was like sending the Cub Scouts in against the Green Bay Packers. "The cans were obscured for the moment by a smokescreen of industry 'word pollution,' " said Rodgers, "but the smoke will clear and the cans will still be there. And as they mount up, it will be harder for industry to hide them behind confusing rhetoric."

The same was true in New York City, where Jerome Kretchmer's EPA crew toiled for months drawing up a comprehensive packaging tax. It was approved all the way up the line, including the State Legislature in Albany, which normally couldn't care less, much less give a damn about NYC's massive urban problems. And where was it defeated? Right in NYC's own city council. A massive campaign was waged by industry along with a lot of back-room arm-bending. And when the smoke-filled rooms cleared, the city council "in all its wisdom passed only the plastics tax," fumed Kretchmer. The plastics industry immediately hauled the city into court and, rightly so, won their case charging discrimination.

As for all the pious pointing to "reclamation" centers across the country, often they are no more than a barrel placed in front of the can company's factory, which is, needless to say, not the most conveniently located reclamation center. Barrels are cheap.

Not that the bottle and can industries do not recognize the hue and cry around them calling for returnables instead of disposables. By the end of 1971 some 204 pieces of restrictive legislation in forty-four states, 29 proposed ordinances in cities and counties, and 14 bills before the U.S. Congress were all aimed directly at cans and bottles.

Judd Alexander says, "Many people *say* they would buy and return these bottles. The facts do not bear this out. Instead of forty to fifty round trips, the average returnable now makes less than four in our major cities." He cited the can industry's favorite example. "In a Detroit survey, 70 percent of the women shoppers said they would buy returnables, but failed to do so when that same supermarket chain put them on sale." The Detroit debacle recurs over and over in papers, speeches, conversations.

(Detroit housewives had their own reply to that: They charged that supermarkets either displayed the returnables badly, sticking them behind cans and disposables, or refused to stock more than a meager minimum supply.)

When WCBS-TV in New York went on the air with an editorial blasting the throwaway bottle and can, one of the defendants was allowed equal time under the fairness doctrine. What WCBS said in their editorial was that throwaway bottles and beverage cans were a major garbage problem and should be recycled. Sounds fair enough. There with another point of view was Abraham M. Raboy, vice-president of the Yoo-Hoo Chocolate Beverage Corporation. It was a typical reaction from industry as Mr. Yoo-Hoo waded into the sticky subject of throwaway bottles. "What about all those other things that come in cans?" the Yoo-Hoo man asked. Baby food and dog food and canned fruits and vegetables and all those other goodies. Yoo-Hoo demanded equal treatment. "If you attack the beer and beverage business, why not every other packaging material?" Fair enough. Besides, he pointed out, "the nonreturnable container is a phenomenon of our pleasant society. The reason the public is getting nonreturnable containers now is because people didn't feel it was worth carrying the bottle back to the store."

Here we are, living in a glass house in a plastic world, threatened with a severe case of metal fatigue. But that's not all. Ah, would that it were: there is paper. Outside, a paper tiger is prowling. Paper for love letters and birth certificates and divorce decrees and shopping lists and mash notes and ransom notes and prisoner demands. Paper for declaring war and accepting surrender. Paper in duplicate and triplicate. Nine million tons of newspapers to read all about all those other papers we used.

And paper for packaging. Now paper is 50 to 60 percent of our municipal garbage, and packaging is 14 percent of that garbage. And paper packaging is 55 percent of the 14 percent. Or: paper accounts for over half of our packaging headaches.

On the one hand, paper is our biggest packaging headache. Excedrin Headache No. 25. Every year some 25 million tons of paper is used in packaging, much of it excess. Corrugated shipping cartons, suit boxes, shopping bags, detergent boxes, boxes for the raisins, boxes for hosiery and shoes and candy and cosmetics, cardboard liners for shirt collars and plastic-packaged panty hose, milk cartons and butter boxes and frozen-food cartons, paper on cigarette packs and around soap, the cardboard that holds flash cubes and batteries and razor blades. Paper around nasturtium seeds and sewing machine needles and . . .

On the other hand, paper is good, if you are something of an ecological Pollyanna. Paper is both recycled and recyclable. Besides, paper is biodegradable: the paper packaging that does make it out to the town dump, if not compacted too much, breaks down of its own free will and, in turn, nourishes the soil with all its cellulose and other natural nutrients.

So paper is not as bad as plastic (which cannot be recycled and which does not break down). But it is bad enough.

Americans consume 57 million tons of paper a year. Conservative estimates figure that paper is about 50 to 60 percent of our municipal garbage load in this country. Dr. Stephen Varro, who buys New York City garbage for his composting plant in Brooklyn, figures NYC's gourmet garbage is upward of 80 to 90 percent paper. Last year 5 million tons of paper were used in business communications alone. The Simpson Lee Paper Company of San Francisco estimates that by the end of this century businessmen can get that load up to 15 million tons if they just Try Harder. In 1970 each person in the United States used 556 pounds of paper products—almost all of it disposable and disposed. One-time use. Userism. That 57 million tons. "The only trouble with consuming 57 million tons of paper," say the people at Simpson Lee, "is in the consuming."

We are fighting a paper tiger that threatens to bury us. He weighs 57 million tons, 25 million of that for packaging. Every day we haul him around with us, then throw him away.

The outlook for paper? Like everything else in packaging, there is nowhere else to go but up. We used 25 million tons in 1966; by 1976 our Userism will be up to some 35 million tons.

All this tonnage is very interesting. Environmentalists figure that for each ton of paper produced, seventeen trees have had to be chopped down. The paper industry itself owns some 50 million acres of commercial forest. Beyond that it and the lumber industry lease ten times that amount from private and federal sources, bringing the total commercial forests in this country to 500 million acres. The plot thickens a bit when it is discovered that of that 500-million-acre total, some 97 million acres of forest land—nearly 20 percent—come from the national forests. Many of us naïvely assumed, that national forests were above such commercial interests as logging and cutting. There they stood—or so we thought—the forests primeval, growing and soaring and towering.

It has been discovered that the national forests have been steadily encroached on by those commercial lumber interests. The federal government, hand-in-glove with commercial interests, has allowed our national forests to be cut down. Supposedly there is a marvelous balance between what is cut down and what remains. The Forest Service has worked up an "allowable cut" program; but the "allowable cut" figure keeps going up and up. In 1950 the "allowable cut" in national forests was 5.6 billion board feet. In 1971—despite a supposedly stable stock of standing timber—the "allowable cut" jumped to 13.75 billion board feet. In addition, the Forest Service sanctioned and stood by while something called the "clear-cut" method of timbering was employed. It is just as the name says: a stand of trees—hundreds of acres—is completely cut away, leaving a raw gash on the landscape. This got environmentalists hopping mad.

"The Forest Service's management policies are wreaking havoc with the environment," charged Wyoming's Senator Gale McGee. "Soil is eroding, reforestation is neglected if not ignored, streams are silting and clear-cutting remains a basic practice."

"How the allowable cut was doubled with no increase in available standing timber or growth rate is no mystery," charged Charles H. Stoddard, former head of the Department of the Interior's Bureau of Land Management. "It is accomplished by computers: by reducing the rotation age of the next crop, by adding 'protection' forest land on steep slopes or poor soils which earlier surveys showed should not be logged."

No matter. The culprit is consumption, and the guy left holding the paper bag is the consumer. Our trees—be they privately owned or part of the national forests—are being cut down and, sadly, often for nothing more than excess packaging. That box around the waxed paper bags holding the potato chips. That box around the individual packets of dumpling mix. That cellophane around the box around the paper around the individual tea bags.

"The Forest Service is a wholly owned subsidiary of the timber industry," charged Representative John Dingell of Michigan.

Not that all the timber cut goes for paper. Nor does it all go for excess packaging. The timber goes into newspapers, magazines, stationery, wood for building homes, making picture frames, and forming educational toys for children. Much of this is justified. What is not justified is many of those 25 million tons of packaging.

Even the paper industry is a gloomy doom-monger in the consumption question. "The increase expected in the nation's population and in its standard of living points to greatly enlarged consumption of paper and paperboard," a spokesman for the American Paper Institute admitted in

1971. Further, the institute estimates that "by 1986 the [paper] industry will need 120 million cords of pulpwood, or about double the amount used in 1969. The problem is deepened by the probability that as federal and state agencies and developing municipalities seek more land for highways, airports, housing developments, wilderness areas and other public uses, inroads will be made on forests, including those owned and leased by paper companies." At an informal ecology meeting in a Manhattan living room early in 1971, a spokesman for the API—after much scooting around in his chair and loosening of his tie and ceiling gazing—figured that yes, maybe we would run out of available timber as early as 1980. It is enough to send us all straight back to writing on both sides of the paper and bundling newspapers and saving old envelopes.

Meanwhile, technology is offstage concocting a little something for our packaging pleasure. As if plastic were not bad enough, as if aluminum were not bad enough, as if we were not already hounded by the excesses of our present packaging, technology has invented the composite can. It is a wonder, combining everything but the kitchen sink. The composite can is a layer of polypropylene, a dash of aluminum, a pinch of polyethylene, a twist of kraft paper, a layer of paperboard, and a topping of aluminum foil. It will be cheaper, stronger, more durable than anything else. And, unlike aluminum, totally nonrecyclable. It would take the Jolly Green Giant working overtime for centuries to come up with the technology to separate out the aluminum that is connected to the polyethylene that is connected to the kraft paper.

Fortunately our incinerators and landfills will not be graced with the presence of the composite can for a few years to come—five at least—but it is there, waiting for us.

There occurred, from September 22 to 24, in 1969, a significant meeting. For it was then that the First National Packaging Wastes Conference was held in San Francisco. Given the fact that we traditionally wait until the problem is well upon us before we, first, recognize it as a problem, and second, begin to do something about it, the First National Packaging Wastes Conference marked not the beginning of the problem, certainly, but the beginning of concern over the problem. Some three hundred delegates attended. Delegates from the packaging industry—metals, plastics, glass, paper—plus scientists, engineers, psychologists, sanitation men, newspapermen.

What marked the conference was not only the concern expressed for garbage in general, but for packaging garbage in particular. Speaker after speaker discoursed on this subject and, generally, speaker after speaker blamed (1) the individual consumer and (2) the packaging industry itself for the burgeoning and booming packaging problem. For if business is booming in the packaging industry, so is the problem booming.

What was also sadly evident at the conference was the defensive cynicism of the packaging industry. Their solution was not to cut down on packaging but to make the consumer pay for it, figuring that since the consumer was the one who created the waste, he should therefore bear the burden of the cost of disposal. No thought or concern was given to the incredible social costs of destroying natural resources. No thought to the taking of our limited supply of natural resources—trees, petroleum, steel—to create this packaging of paper, glass, aluminum, steel, plastic. No thought to the high actual cost of manufacturing it (which, of course, is translated into "profit" for the manufacturers), or to the ultimate cost of disposing of it. Industry insisted that since consumers demand increasing convenience, service, and attractiveness in their purchases, industry has only responded to their demands. The industry spokesmen rarely listen to the argument that if that is all there is to buy —double-packaged, excessively packaged goods—what is the poor consumer to do? Besides stand at the check-out counter, that is, and patiently tear off as much excess packaging as she can, to protest her ultimate rip-off at the hands of the packaging industry.

"Though we have landed a man on the moon," said Representative Henry Reuss of Wisconsin, "we continue to make a wasteland of the earth. In plain words, the Federal Government is defaulting on conservation. While subsidies to large corporate farmers to grow unneeded crops, spending on military and wasteful public works projects go largely unchecked, vital environmental programs are increasingly short-changed." Reuss's suggestion: "Let's ease up on space exploration for a while and explore ways to make the earth worth staying on."

Irving Bengelsdorf, science editor of the Los Angeles *Times*, pointed out that the packaging industry was on the one hand howling with outrage at the mere thought of having to pay some sort of packaging taxes (which they promised would be passed on to the consumer, thus leaving the industry holding not the bag but a money sack) while on the other demanding government financing for basic research in the packaging field. (About this time Congressman Henry Reuss waded in and pointed out that this country now spends only $15 million on waste disposal research as compared to over $300 million on chemical and biological warfare research.)

But perhaps Leonard Stefanelli, president of the Sunset Scavenger Company, the people responsible for picking up San Francisco's garbage, puts it best: "The packaging industry has created one hell of a situation. The United States being one of the most productive countries in the world has charged forward in the development of more beautiful packaging to attract the consumers but has also created another problem of what to do with these 'beautiful products' after they have been used only once." As Stefanelli argues, "You [packaging] people have made my job a tremendously more difficult one to perform."

So far the best, most logical, most cogent, and most efficient answer to the whole packaging problem—plastics, paper, glass, aluminum—came from, of all places, the U.S. government, which suggested that we:

1. Reduce the quantity of packaging materials used, thereby reducing the quantity of such wastes that must be transported and handled.

2. Reduce the destruction of valuable natural resources.

3. Reduce the technical difficulty of handling such wastes in disposal or salvage facilities.

4. Dispose of solid wastes more effectively and efficiently by known methods such as landfill and incineration and by new approaches to solid waste processing.

For if we cut down on packaging, we save trees and petroleum and metals and silica. We lower the risk of oil spills and strip-mining desecration. Then, in the end, we cut down on our tremendous and overpowering need for more landfill space, thus again saving our natural resources —in this instance our very physical environment: land.

Somehow, carrying home an eggplant *au naturel* seems worth it all.

V

Good Riddance

Who *cares* about garbage? Precious few of us. We burn it in our disgraceful incinerators (75 percent of which are deemed inadequate by the good gray government) or we bury it in dumps (94 percent of those also being designated as inadequate). Is this any way to treat our garbage? If we can carefully and meticulously tear down the Metropolitan Opera House in New York, a gargantuan structure of red brick and memories, can we not with the same care and concern dispose of our garbage? One would surely think so. After all, this is historic tradition we are dealing with. America has showed she could efficiently rid herself of Pennsylvania Station—a chore that took years and cost millions—which came replete with five-story marble columns and more eagles and gargoyles than could possibly be counted, not to mention colonnades and soaring arches and murals on the ceilings. Why can we not apply the same devotion and care to the destruction of our garbage? The tender, loving care with which we handle the demolition, destruction, and disposal of the likes of the Met and Penn Station does not, unfortunately, carry over into our orange peelings and Saran wrap and Pampers and beer cans and aluminum trays and plastic detergent bottles.

There have been modest attempts to clean up our dumps. Kenilworth, an area of Washington, D.C., for years operated a dump that was a wonder to behold. It spread itself generously over the district's natural environment, an artful collection of residential residue from the leadership of this country. Here a senator's necktie, there a representative's shoe. An oatmeal box from some diplomat. A beer can from the Kitchen Cabinet. Every afternoon at an appointed hour, the origin of which has been lost in the mists and clouds of garbage history, fires would be lit in the dump. Beacons, guiding the "pickers" who were allowed to roam through that gloaming of governmental garbage gleaning a few treats and treasures. Alas, one day two young residents of the area—which, by the way, was short on playground space—were caught between the cross fires, as it were. One fleet-footed seven-year-old escaped the fanciful flames; the other perished. Folks didn't take too kindly to the Kenilworth dump after that, and it was finally cleaned up and turned into a sanitary landfill.

Periodically some shards of imagination poke through the garbage that has piled up around us. Some farsighted folks in the garbage business looked at garbage, looked at the dump where it was deposited, then shut their eyes and rocked back on their heels and put their heads to work. From dumps to sanitary landfills to land. Something from nothing. (Well—if you consider wetlands and shorelines and natural holes in the earth to be nothing, that is.) And so it came to pass, out of the darkness of at least some of the despair over our garbage, that parks and playgrounds and airports and recreation areas blossomed forth on the face of our land. The idea began to catch on, for behold, in the east, where garbage piled up faster than anywhere else, it was discovered that something could be created from nothing.

In short: pile that garbage deep enough and thick enough and heavy enough, and you've got something pretty substantial. LaGuardia Airport serving the New York City area is made mainly of landfill. The magnificent Tidal Basin in Washington, D.C., is landfill. Montebello, California, collected its garbage for years in a confined area and then, when the final beer can hit its mark, the whole thing was turned into a police-sponsored playground. San Francisco (which was prohibited from dumping raw garbage into San Francisco Bay in 1965, something it had been doing for years) turned its natural tidelands into parking facilities and stables for the Golden Gate Track by dumping its garbage there. Some useless old sandpits outside Dallas were turned into a parking lot and a park. Richmond, Virginia, built shopping center parking lots out of sanitary landfill. (Richmond also built houses on completed landfill, then ran into a most troublesome problem: the landfill had not settled enough before construction was begun, and what to the homeowners' wondering eyes should appear but cracks in their foundations.)

Right next to the University of Washington campus in Seattle was a steaming, fetid, dank swamp that supported all manner of unrecognizable flora and microscopic fauna in its 166 acres. Back in 1933 it was decided the land could be reclaimed. When landfill speculators set to work they discovered the damn thing went down to a depth of 60 feet in some places and generally had the consistency of undercooked pudding. Nobody knows exactly how much garbage was donated to fill in the 166 acres, but in a ten-year period in the 1950s and 1960s over 8 million cubic yards of garbage was deposited in the swamp. Dikes were built, drains and canals constructed. Today the swamp is a combination playfield, driving range, parking lot, and athletic field house.

Another ingenious landfill is up in Niagara Falls, New York, site of a new garbage dump. Seems there was an ugly canal just sitting there, a blight on the landscape. The old canal, once used for hydraulic power, measured 90 feet wide and 40 feet deep and 4,440 feet long. And it stretched right smack through the Niagara Falls business district: wet, damp, unsightly. People threw the ubiquitous beer can into it, then kept it company with sandwich wrappings and old shoes and hangers and plastic cleaner bags and prophylactics. The canal was drained and a scientific program of sanitary landfill was instituted. A clean layer of junk was put down, compacted, and covered by dirt. Then another compacted layer of random garbage was put in. More dirt added. (Making a sanitary landfill is very much like making lasagna: you work in layers. A layer of garbage, a layer of dirt, a layer of garbage, a layer of dirt. Ending up with a layer of dirt instead of mozzarella.) When it was finished, the city fathers used it for business expansion and highways.

The landfill list goes on. In Chicago the fourteen-story high-rise Winston Towers apartments are built on 55 acres of landfill. John Sexton Contractors utilized fifty-foot pilings driven through the landfill and down to bedrock to support the $35 million apartment complex.

Flushing Meadows, New York—as a result of the 1939 World's Fair—is 1,200 acres of landfill. It was used as a dump for years, then reclaimed specifically for the World's Fair. Today it is park and residential property. The Food Distribution Center in Philadelphia, a multi-million dollar complex of 36 buildings, sprawls over 338 acres that once were a jungle of foul trash-burning dumps.

In the area surrounding San Diego are hundreds of small canyons that undulate over the landscape. If they were noticed at all it was by land speculators and contractors who brought the best of engineering science to bear on the problem of turning them into tract housing developments. But now science and technology are truly catching up with those canyons: under a grant from the federal government, the city of San Diego is eyeing them as resting places for garbage. One site is already being worked over. Its final goal: an archery range and a golf course.

There are problems with the landfill method of garbage disposal, however. Witness those houses in Richmond, Virginia, with cracks in their foundations. Landfill is a veritable beehive of activity, for garbage is not inert. In a series of studies done on New York City landfill, temperatures were found to reach dizzying heights—160° at times—from the microorganisms that were munching away in the garbage. (That was very unusual, but then all of New York City, including its garbage, is unusual. Generally landfill temperatures hit around 100 to 120°.) Additionally, there is a lot of productivity of new natural resources going on down there in the depths of a landfill, sanitary or not. One of the most prevalent of resources being constantly created in landfill is methane gas.

Seattle discovered that the best way to get rid of its landfill odors was to burn off the methane gas. Flames reached eight to ten feet in the air. Workmen in another landfill were even more inventive: they drove a one-inch pipe down into a particularly ripe section of the landfill, fashioned a tip on the end of it, and fired the gas that rushed up the pipe. It provided a continuous, steady, strong flame for making coffee. (This is also a danger when constructing atop a landfill. According to the American Public Works Association, "Gases have seeped out of the fill and upward along an insulated water or sewer pipe from the fill to an enclosed space in the building. A spark from an electric fixture is all that is needed to ignite the gas.")

Perhaps the most sensible idea now taking form—both to environmentalists and to garbage men alike—is the scheme to use garbage as fill for abandoned strip mines. The country is scarred from coast to coast by these mines, whose raw wounds fester and erode the landscape. West Virginia, Pennsylvania, Washington. Until recently, when governmental and environmental pressures were brought to bear, the mines were gouged out, used, and abandoned. Only in recent and rare cases are they replanted, once the topographical and ecological damage has been done. New York City once had a scheme to rail-haul its garbage to Pennsylvania, then sell it to landowners whose property contained abandoned coal mines. The plan, as usual, was fought from start to finish by states and municipalities who do not want somebody else's garbage. When Joseph M. Missavage, Broome County, New York, planning director, mused aloud about sending his area's garbage off to Pennsylvania and depositing it in abandoned coal mines, he got a reaction to his idea from Pennsylvania officials: It stinks. Missavage figured the plan would not only help get rid of garbage and abandoned mines but would provide work for unemployed Appalachian miners.

Washington State, however, in mid-1971 proposed rail-hauling its refuse to abandoned mines out in Lewis County. At the urging of both Governor Dan Evans and the state's Department of Ecology, the plan would dispose of a majority of the solid waste generated in the central-southeastern sector of the state, plus portions of Oregon. The destination would be in the environs of Centralia, Washington, where an excavated strip coal mining area is located. Observers calculate that the 20,000 acres available for reclamation would handle all the waste from the 2.6 million people living in the proposed service area for at least thirty-five years. It is indeed a step in the right direction. Although current proposed legislation—specifically a bill introduced in the House by Representative Ken Hechler—would prohibit any new strip mining, the country is still suffering from the plunderful life strip mining has enjoyed so far.

There is one thing wrong with the idea of strip mines as the final glorious resting place for our garbage. At least three states—Delaware, Maine, and Rhode Island—have laws against importing garbage and others have similar legislation pending.

G. H. Hamlin, writing in the *Refuse Removal Journal* back in 1967, proposed rail-hauling San Francisco's garbage out into the desert and disposing of it there. Hamlin figured filling in to a depth of 18 feet over 9,125 acres would give a landfill that would last for 50 years. This far-sighted approach (well, at least as far as the deserts outside San Francisco) was echoed by the private contractor who serves the whole San Francisco peninsula—two million people—with garbage pickup, salvage, and disposal.

One of the knottiest natural problems facing the agricultural states in this country is that of soil erosion. To combine two natural resources—garbage and gullies—the Department of Health, Education, and Welfare found a gully in Sarpy County, Nebraska. Located some ten miles south of Omaha, this demonstration gully was just what the government was looking for. They cleared it of brush and trees, built an access road to it, fenced it in, built an administration hut, and started loading it with garbage. The gully, 4.5 acres of unfarmable land, was in operation for eighteen months, taking twenty-five tons of garbage a day from the surrounding towns and villages. When the eighteen months were up, those 4.5 unfarmable acres were leveled off and seeded to bromegrass for erosion control. Eventually the area will be planted with corn, and it is expected it will increase the landowner's net income by some $225 per year. The government also figured that the value of the land will rise from $5 a worthless acre to $505 for the new reclaimed land.

That was just a 4.5-acre gully in Nebraska. Some would consider it absolutely nowhere as far as garbage is concerned. The real excitement concerning garbage landfill is in Virginia Beach, Virginia. Virginia proudly puts it right out there in the open: they are building a mountain out of their garbage. Mount Trashmore. Since April of 1967 the residents of Virginia Beach have been donating all the fruits of their labors to the construction of their garbage mountain. Citizens donate, personnel from Old Dominion College monitor the whole shebang, the federal government kicks in some cash. By 1969 the height had reached fourteen feet. By 1972 it was a hundred feet long, seventy-two feet high, and rising.

When Mount Trashmore is finally filled in, it will cover a total of 161 acres. The plans are grandiose, including a 10,000-seat amphitheater (the mound for this has already taken shape and its outlines are already rising out of the rubbish). There will be a theater, a concert dome, and parking facilities, plus a soap box derby course, a baseball field, tennis courts, picnic areas, and an all-purpose playing field for football, soccer, and just plain chasing around.

Mount Trashmore has its temperature taken regularly, and so far it is a nice healthy mountain with nary a trace of excess gas in its system. Bill Beck, one of the observers from Old Dominion, reports no rats have been sighted and there is no problem with flies. (One thing, though: sea gulls. They swoop in to what has probably come to be known in the gull underground as the gourmet restaurant for the Eastern Seaboard. Thousands of them fly in every day to the annoyance of the workers. Not that American technology is not up to sea gulls: the workers set off anti-gull firecrackers to scare the gulls away. It works. Beck is nothing less than enthusiastic about the place. "Not only will it work," he says, "but it will be one of the most beautiful spots in Virginia Beach. Just wait'll we get it spruced up and grass planted." Beck admits to a personal bias, however. After all, "When you've lived with something for three and a half years," he said late in 1970, "you're bound to have a personal feeling for it."

The Mount Trashmore approach to garbage is not all that new. At the end of the war Berlin found itself with its own unique kind of garbage: bomb rubble. By war's end the demolition debris was estimated at 80 million cubic meters and was piled all around the city. What to do with it? In a spirit of cooperation the Germans decided to give it back to its creators. The French were presented with a big pile of bomb rubble, dumped in a huge bunker in the French sector of Berlin. Once the bunker was filled in, dirt was spread over it and trees and grass were planted to make it a public park. The next deposit was made in the American sector. Appropriately enough it soared into the air with a swimming pool at its base and an observation tower atop its summit.

But the biggest of them all is Rubble Mountain, as the Berliners call it. Some 25 million cubic meters of debris were donated to this spot. It rises 360 feet into the air, has grapevines growing on one side, an observation post on its summit, and ski and toboggan runs. It took ten years to build to its present height and, ironically enough, stands on the site of a former Nazi military college.

For the most part, though, we dump our garbage. Its ignominious final resting place is just as its name implies: a dump. We dump it in and then verbally dump on it. It breeds rats and mice and mosquitoes and provides bad habits for the bears in Yellowstone and little boys in Kenilworth. We throw it around willy-nilly with never a care for it aesthetically. Our garbage is a graceless entity indeed.

There is, however, a dump in Churchill, Manitoba, that was much loved by polar bears who liked to rummage through the rubbish. Townspeople began getting irritated when the bears began strolling brazenly through town, periodically peering through windows and otherwise disturbing mealtimes, naptimes, and TV watching. It is hard, even in Churchill, to watch the six o'clock news with a polar bear peering in the picture window. (The Royal Mounted Police were even called in one night when a bear poked his head through the window—when it was closed —of the Reg Lacey house while Reg and his family were trying to eat dinner. A shoot-out ensued between the Mounties and the unarmed bear.) An airlift was forthwith organized. An old DC-3 plane, equipped with cages, was hauled into service. Townspeople, spearheaded by conservationist Brian Davies, executive director of the International Fund for Animal Welfare, loaded all twenty-four town polars onto the DC-3 and flew them 150 miles away— at $40 a head—to what they figured would be equally happy hunting grounds. Fifteen days later, two of the bears were back. The rest began drifting in one by one. And since polar bears normally lumber along at the rate of about five miles a day, wildlife men in the Churchill region figured the rewards of the Churchill garbage dump must have stimulated the bears' natural homing instinct.

Scottsdale, Arizona, is something of a dream town. On the surface it looks like any ordinary Hollywood movie set, except that no bad guys come roaring out of McGee's Trading Post. Just tourists, armed with Navajo rugs and Hopi baskets. There is a sort of somnambulant Social Security quality about Scottsdale, redolent with cowboy boots and gingham dresses and soda shoppes and cactus plants. The streets are wide, the buildings are low, the sun is lemony yellow and the sky is not cloudy all day. But looks can be deceiving. Beneath that calm exterior beats the heart of a tortured city. The sky might not be cloudy all day, but all over town storm clouds were gathering. Scottsdale was a town threatened with garbage.

Once confronted with the problem, the Scottsdale town fathers knit their brows as one, joining together in the good old American way to solve their problem. Get a little government money here, work up a demonstration to study there, and before you could say go home, garbage—the problem was solved.

Scottsdale could not be prouder of its invention, for indeed it is an invention. They call it, proudly and publicly, "Son of Godzilla." But first, a little historic perspective, for if there is a son there has to be a mother, right? Indeed, there was a Godzilla. She was something of a Bride of Frankenstein, a makeshift creature formed out of old parts and bits and pieces. Essentially she was a front-loading garbage truck to test the idea that a truck could mechanically reach over and pick up the garbage in question and throw it over its shoulder into its hopper. "Godzilla has performed well," the Scottsdale Department of Public Works said of their mechanical mother. "She was not built for speed or efficiency, but only to do a job that couldn't be done by hand." Godzilla was retired from the fray with thanks for a good job well done and the promise that she will be "retained as a standby unit."

From Godzilla certain basic truths were learned. These truths (sometimes called "mistakes") were applied to what is known today as "Son of Godzilla." A mechanical garbage truck, Son of Godzilla is truly wondrous. He features an air-conditioned cab with a cassette tape deck and an AM-FM radio. The operator sits in there, stoking up on Mantovani and picking up the garbage. This is done by a big arm that reaches out from the front of the truck, picks up the containerized garbage by its fingers, lifts it up and dumps it over its shoulder into a hole in its back. Son of Godzilla even reaches around cars. Scottsdale's Public Works Department proudly points out that Son of Godzilla provides twice-a-week service for "less than one dollar per home per month." Each home is equipped with a big (80-gallon) container, which they place in front of their home, or in the alley if that is where pickup is, and into which they dump their garbage. Simple. Then Son of Godzilla comes along and empties it, puts it back down, takes the garbage off with him. He can service 1,500 homes a day.

Beyond the wonders of young Godzilla himself, the Scottsdale project also figures that he has good side benefits, too. "Physical fitness is no longer a limiting prerequisite. Older men, physically handicapped, or women may easily be trained in the simple operation of this equipment." Just sit there with your tape deck and necktie and let young Godzilla do the work.

Nifty as Godzilla is, he is definitely a small fry. Scottsdale's population is only 10,026 and he doesn't work all the neighborhoods. (Not yet anyway.) And the good folks of Scottsdale, tourists included, only manufacture 3,100 tons of garbage a month anyway. What do the big cities do? Chicago now has a new incinerator operating that its engineers, Metcalf & Eddy, proudly call "the largest now operating in the Western Hemisphere." The Chicago Northwest Incinerator is a behemoth structure covering eleven acres, which cost $23 million to build. It sits in its own industrial park surrounded by grass and bushes and shrubs and 300 parking slots. Although to the innocent eye it looks like just another industrial institution—flat and brick and smokestacks and ugly—to the garbage men of Chicago it is a glory of gargantuan proportions. One of four municipal inciner-

ators, it services Chicago's 3.5 million residents, who produce 9,000 tons of garbage a day. The new Chicago Northwest Incinerator handles 1,600 tons of the 9,000 total. Additionally, this incinerator is capable of producing 440,000
pounds of steam per hour, enough to service itself and
keeps its own pistons and motors and generators and what-
have-you running, plus have some left over to sell off to
surrounding industries.

Out in Franklin, Ohio, a small town of 10,000 nestled in
the Ohio Valley, there is a $2 million experimental recy-
cling plant at work. The citizens of Franklin merely donate
their raw garbage, and the plant goes to work. Built by the
Black Clawson Co., this little number now handles 50 tons
a day, a mere drop in the garbage bucket compared to, say,
the 29,000 tons New York City produces each day. First of
all, insert your garbage. Plain ordinary right-from-the-can
garbage. The magic of the Black Clawson machine takes
over immediately. A conveyor belt takes your garbage into
the Hydrapulper, where it is soaked with water. Then a big
rotor comes in—think of a great big eggbeater—and
mashes the whole mess up into what the garbage guys call
"liquid slurry." It looks very much like runny oatmeal.
Things like bedsprings and tire irons and heavy metals are
ejected at the side, leaving only this slurry made up of
paper, food wastes, plastics, leaves, glass, and bones. The
slurry is then taken along to a magnetic separator, which
cleans out those smaller chunks of metals for recycling.
Then the slurry is pumped into a cyclone where it spins
around (think of the final spin-dry cycle on the basement
washing machine) and centrifugal force separates out the
heavier materials—glass, dirt, sand—which flops to the bottom. The rest of the slurry, by this time composed mainly
of food, plastics, and paper, gets sent along to something

called the "Fiberclaim system," where the paper fiber is removed for recycling by running the whole mess over a series of screens which somehow magically separates the paper fibers from the rest of the gook. Finally the residue is dewatered and dried into a cakelike substance that is in turn broken into lumps and pellets (⅜ to 1½ inches in diameter) that are whisked pneumatically to their final disposal: being destroyed at 1500° F. temperatures. It is incredible.

The garbage is reduced by 90 to 95 percent in volume and 75 percent in weight. Metals and paper have been salvaged for recycling. Right now, the Glass Container Manufacturing Institute and the Sortex Company of North America have developed a process to grab the glass as it floats by, thus enabling it to be recycled too.

It costs $6 per ton. Early in 1972 the Black Clawson process was recycling out five tons of metals and ten tons of paper a day. Experts figure the glass will hit three tons a day. Are the project people excited about their garbage machine? "The plant is performing beyond all our expectations in processing the refuse," says N. Thomas Neff, project manager of the complex.

Beyond that, the possibility of energy and heat recovery from the process is a big plus factor. The combustible refuse left after separation has a potential heat value of about 7000 BTU per pound. Recovered heat could be used either for fuel or for power energy.

The ceremonies in Franklin marking the dedication and opening of their new garbage plant will go down in that town's history as a truly all-American day. American Legion Post No. 149 attended to the Flag Raising Ceremonies. Miss Phyllis Darragh sang "Our National Anthem," backed up by the Franklin High School Band (John D. Palmer, Director). Then All joined in for the Pledge of

Allegiance. There was a Welcome and the Presentation of
City Key to Mr. Richard Vaughn.[1] Then "Pressing the
Starter Button" starring Mr. Vaughn. Afterward, thousands of Franklinites trooped through the big facility for
the Public Inspection. As the plaque says, it was "Dedicated to the citizens of this small community who had the
foresight and courage to save the purity of the land en-
trusted to them by God."

Franklin is proud of its garbage facility—and rightly so.
"Franklin Dedicates World's First Garbage/Trash Recycling Plant," sang the banner headline in the August 11
Franklin *Chronicle*. "Every resident of this comparatively
small industrial city simply by taking out the garbage, now
practices total recycling of household refuse." It was
enough to make an ecological doom-monger sit right down
and recycle some organic tears.

Another new process looming large over the landfill is
Monsanto's "Landgard" process. It is a very hard-nosed,
pragmatic approach to garbage: get rid of it. "There has
been a lot of garbage generated about garbage," says Curtis
Snow of Monsanto. "Recycling and recovery holds a lot of
fascination for politicians and the lay mind," he says. But
to him that doesn't answer the question of what you do
with the stuff once you've recovered it from the garbage
heap. Working from that point of view, the Landgard system just burns the stuff up in the best and most efficient
way: pyrolysis, a high heat/no oxygen burning process. As
Snow poses the question: "What good does it do to recover
something from solid waste if you can't find a home for it
in the economy?" This is not to say resource recovery is out
of the question: the Landgard system could be recon-

---

1. Richard Vaughn: then Deputy Assistant Administrator for Solid Waste Management Programs of the Environmental Protection Agency.

stituted with existing technology to salvage metals and glass. But "we want paper—it burns!"

If the Black Clawson method is a marvel of a Rube Goldberg machine that mixes and separates and shoots metals off here and paper off there and garbage down there, the Landgard is as simple as an instant cake: just pop the garbage in the mixer and cut it all down to size (about three to four inches in diameter). Then put it in the oven and bake it at a very high temperature without any oxygen. Pyrolysis means, simply, "destructive distillation, carbonization," which is, as everyone knows, "thermal decomposition in the absence of air." Once the garbage has been baked it is run through some water to cool it, and then it is trucked off to the friendly neighborhood landfill. Volume reduction: 94 per cent. Cost: about $7.50 to $8.50 per ton.

Alas, neither the Black Clawson method nor Monsanto's Landgard process have been tested on any large commercial scale. Black Clawson handles 50 tons a day in Franklin (aiming for 150 soon), while Landgard was managing 35 tons a day on an experimental basis. Both promise they can deliver, even offering municipalities intriguing deals: we'll build the plant, you lease it from us. Turn-key operations. Late in 1972 New York City was considering a 1,000-ton-per-day Landgard system with the possibility of small-scale resource recovery, provided it could get seed money from a recently passed Environmental Bond Issue.

Landgard's pyrolysis and Black Clawson's drowning method are mild compared to the Heil-Gonard process, which is nothing short of garbage masochism. It starts out simply enough: insert garbage. Then two hammer mills, as they are called in the trade, take over and pummel the garbage beyond recognition. Godzilla wouldn't even recognize this mess. One of the two hammer mills operates on the "horizontal grinding and ballistic rejection" system, which is as its name implies: the big hand rubs the garbage horizontally. The second hammer mill does the opposite: it operates under the "vertical grinding and ballistic rejection" theory. It is the Mickey Spillane approach to garbage: kick the shit out of it. Once done, the garbage is disposed of in sanitary landfill where all sorts of benefits begin making themselves immediately felt: it doesn't smell as much as regular raw garbage. It settles better and faster, and therefore a municipality doesn't have to wait years to build that airport or that park or that shopping center on it. Finally, it doesn't blow up as much. (Dumps and/or landfills have a tendency to ignite and blow up periodically. This not only scares the neighbors out of their wits but fries up a sizable rat population at the same time.) Tests have shown that the so-called Seattle Torch could never occur where garbage has been beaten into submission by a Heil process. But—more and better—rats don't stand a chance. Along with polar bears and grizzlies, most dumps are happy hunting grounds for rats. Rats cannot survive on a Heil diet. Purdue University ran a little test serving only Heil-processed garbage to their laboratory rats, who cannibalized themselves within sixteen days.

The first commercial application of the Heil system came with the opening of a pummeling plant in Pompano Beach, Florida. The two-story structure housing the operation is a blue and white building nestling in a landscaped "oasis" complete with palm trees, lawn area, and a small pond with its very own fountain and live alligator. Very Florida. As in an experimental Milwaukee plant, paper and metals are separated out for salvage. Said Florida Governor Reubin Askew proudly at the plant's formal dedication ceremonies in September of 1971, "This project, which seeks to turn a problem into an opportunity, illustrates that private industry can utilize its technical skills to benefit society." The Florida operation is currently buzzing along far beyond both the original great expectations and the original design capacity: it processes about 225 tons of garbage a day, thus exceeding its 15-ton-per-hour limit by about 6 to 8 tons.

General Electric has a new incinerator design that it is now testing out in Shelbyville, Indiana. The GE people call it the "Vorcinerator" and it operates on the "tornado on its side" principle. In the "tornado" principle the garbage is whirled around in the incinerator in a vortex as it is burned at temperatures of 1800° to 2000° F. The incinerator looks very much like a large drum that has been dropped on its side. Plans call for the garbage first to be shredded, then conveyed to the incinerator, where it is burned at very high temperatures, and reduced in volume and in weight, with the residue being disposed of in sanitary landfill.

The first experimental Vorcinerator successfully handled 1.5 tons of garbage an hour (on an eight-hour day) for Shelbyville and the surrounding area, population 35,000. Phase II calls for a larger Vorcinerator, capable of handling 6 or 7 tons per hour. "If Vortex II operates with the efficiency and economy indicated in the thorough testing of Vortex I, the system will have great potential not only for Shelbyville-size communities but in large cities as well," say the GE people. According to Robert Hasselbring, manager of the project for GE, "Cities won't have to have one, huge incinerator but perhaps five or six smaller ones like this, located in various parts of the city."

Out in Menlo Park, California, the Combustion Power Company is busy working on something they call the "CPU-400," which won't be operable until 1975 or 1976. This, too, is a high-heat incinerator, powered by a gas turbine. Projected capabilities of the CPU-400 are disposal of 400 tons of solid waste per day "in a pollution-free manner" (they *all* promise that) and the generation of 1500 KW of electrical power. According to the Combustion Power people, their CPU-400 will serve communities in the 150,000 to 200,000 population range, "and since the units will be clean, compact and quiet, they can be dispersed about the city to substantially reduce hauling costs." The neighborhood control of garbage again. So far, the CPU-400 is being tested on a one-tenth scale. Quite oversimply, it involves (1) shredding the garbage, (2) air classification of the shredded material, (3) combustion and burning, (4) disposal. Air classification means that after the garbage has been shredded, it is separated into salvageable materials (glass and metals). "We use a large blower to suck the refuse up a vertical tube and force the refuse through a treacherous path before it enters this vertical tube," says Richard G. Reese, General Manager of Combustion Power. The velocity of the air separates out those heavy glass and metal materials and drops them out of the refuse. The refuse continues on through step 3, destruction through incineration, and 4, disposal. When the drop-out residue is collected, the CPU folks, since they have no commercial interest in it, plan to "deliver the residue from the air classifier to some other company that is in the business of separating the metals from the glass, etc."

There is just one thing wrong with all of these highfalutin disposal methods: with the possible exception of the Heil-Gonard system in Pompano Beach and the Chicago Northwest Incinerator, none of them work. There is not a Landgard cum Black Clawson cum CPU-400 among them that works on any practical, everyday, large-scale, long-term scale. With municipalities producing more and more garbage every day and finding themselves confronted with fewer and fewer places to put it, either in the air as incinerated contaminants or in the ground as landfill, American technology is falling down on the job. And if environmentalists are to be believed—and why not? the dismal statistics bear them out—technology will soon find itself not only falling down, but buried in the top layer of an ever growing landfill.

Despite some disclaimers to the contrary, free enterprise is still at work in America. Take the case of John Hollowell, who owns a small company called Hollowell Engineering, Inc., in Detroit. In 1969 John Hollowell's company did $3.5 million worth of business, operated three plants, and had 180 employees doing design work for the automobile industry. But what with governmental pressure for more pollution control and more safety devices rather than more tail fins and more wrap-around chrome, business fell off for John Hollowell. Two years later, Hollowell was down to one plant with 20 employees. "We've been dead for a year," said Hollowell, who had been known to cook spaghetti in the office for his dwindling supply of employees.

But sitting back in the back room of Hollowell Engineering, Inc., behind the hot plate and the spaghetti pans and those 20 employees is, perhaps, Hollowell's Salvation. Or Hollowell's Folly. As Jerry Flint, a reporter for *The New York Times*, noted, the thing looks like a cross between a golf cart and a baby elephant. It's a trash collector, a one-man machine, which scoots around the landscape, sucking up trash with its trunk like a super-sized vacuum cleaner. It operates on the vacuum principle, and the trash that is sucked up through its enormous nostril is whisked back into its body and lands in plastic bags for disposal. Hollowell figures his elephantine invention would go for about $2,000 on the market as opposed to alternative methods of collecting litter and debris, i.e., a man with a stick or a man with a $50,000 machine. "It will work," he says. "It's based on good design principles." The biggest problem: figuring out how to keep the plastic bag from blowing out. "I'm not trying to save the country," Hollowell says. "I'm just trying to make a buck." Well, if that isn't the good old American way, what is?

Perkins Township is a small community of some 12,000 garbage producers that sits half-way between Toledo and

Columbus, Ohio. They may get what can only be considered the truly all-American approach to garbage. If a proposed plan by David Yeager of the Yeager Sani-Pac company of Sandusky goes through, Perkins residents can choose from three methods of garbage disposal: (1) By paying a fee, their garbage will be picked up for them. Or (2) they can drive their own garbage to the landfill site. Or they can (3) take their garbage to what will be the country's, even the world's, first drive-in garbage disposal. If the consumer feels his garbage needs the personal touch—not handled by strangers posing as sanitation men—but the landfill is too far away, there's always the drive-in.

As things stand now, according to Dave Yeager, "a person accumulates a load of garbage and then—provided you get off work early enough—you drive many miles over superhighways to a landfill where you are met with rutted roads strewn with glass and other debris. Then you back up to a big hole—next to a garbage truck—and dump your trash with the wind blowing ashes and paper all over you." Definitely not the way to spend the cocktail hour. "Wouldn't it be nicer," he asks, "to drive maybe five miles or so to some point in the city into an attractive building and have an attendant help you? Maybe even clean your garbage can with an automatic can washer?"

Yeager is the Kurt Vonnegut of garbage, forever envisioning a better world off there somewhere through time and place and space. "Look at the number of gas stations that offer a free car wash with a fill-up. How about free dumping instead? How about the drive-in/carry-out where you go for groceries or beverages? Why not leave a bag of garbage?" After all, he figures, the modern compactor provides sanitary, odorless disposal. Dave Yeager is hung up on those car washes, though. "Look at those 25¢ washes in every city. How many dirty cars do you think there'd be if you had to drive twenty miles to have it washed?" He

answers himself: "There would probably be as many dirty
cars as there are dirty backyards," he figures.

David Yeager, who with his father owns Yeager Sani-
Pac, is truly representative of the young, fresh wind that
is blowing through our garbage today. He obviously believes in revolutionary approaches coupled with the personal touch. For what could be more personally touching
than hauling your very own garbage—garbage produced so
lovingly in your very own home—to the drive-in disposal.
(Young Yeager is also something of an altruist. He claims
to have gotten the idea after having seen cars stuck in the
mud while trying to dump their garbage at the landfill. The
idea of frail housewives trudging through the sludge—
ashes and debris blowing over them—to deposit their garbage in the landfill was, to say the least, aesthetically unpleasant to him.)

"Everybody wants us to pick it up—but nobody wants
us to put it down," is New York City's Jerome Kretchmer's
favorite way of expressing the prevalent American attitude
toward garbage. And in some cases people don't even want
it picked up. Or, at least, so many difficulties are presented
one would think that were the case. Huge municipal and
commercial establishments stagger even the visionary
imaginations of the best of the garbage freaks among us.
Thousand-bed hospitals, five-thousand-unit apartment
houses: how to pick the damned stuff up, not to mention
putting it down.

There is, on the garbage horizon, the concept of pneumatic tubes whisking garbage through hidden underground pipes to a central location where it will then be
picked up. Eastern Cyclone Industries, of Fairfield, New
Jersey, for example, has been in the pneumatic conveyance
business for twenty-five years, handling primarily the
chuting of soiled hospital linens to a central collection

point. In 1968 ECI installed its first Air-Flyte system for garbage purposes in the Alta Bates Community Hospital in Berkeley, California, using the same principle: dump the garbage in handy chutes spotted all over the hospital and it is conveyed to a central pickup point. Currently ECI is making a feasibility study to determine whether this form of garbage manipulation will work in New York City's massive Welfare Island project (5,000 units) and a large multi-unit housing complex for the University of Minnesota campus at Minneapolis. "The principle of pneumatic conveying is simplicity itself," says William Boon, president of Eastern Cyclone Industries. "Basically, it is the drawing of materials by negative air pressure through a conduit from a depository to a collection point." A generator creates the negative air pressure, and bags of garbage weighing fifty pounds or so can be sucked through the tube at sixty miles an hour, up, down, at an angle. They can even turn corners.

Not that this is anything new under the garbage sun. Sundbyberg, Sweden, a fifty-year-old town of some 30,000 persons on the northwestern edge of Stockholm, has a "Centralsug" pneumatic disposal system. Householders simply drop their garbage—everything from neat bags to old Christmas trees, steam irons, umbrellas, indeed anything that can be pushed through the opening—into chutes recessed into their apartment walls. Instead of flopping down to an incinerator where it is burned, the miscellaneous garbage is whisked through those magical underground tubes to a large incinerator. A sifting process lifts out the glass and metals and the remaining refuse goes into the incinerator, where the heat generated is used to provide heat for the 1,100 apartment units now using the vacuum disposal system. Eight such Centralsug systems are now operating, and twelve more are under contract throughout Europe.

Environgenics Co., a division of General Tire & Rubber's subsidiary Aerojet-General Corporation, is the U.S. licensee for the Centralsug system, which they market in the U.S. under the name AVAC. The first AVAC installation is a mammoth garbage unit at Walt Disney World in Florida. Walt Disney World has fifteen garbage collection points scattered over its 2,500 fantasy-land acres. More than a mile of underground piping runs from the Contemporary Resort Hotel around the Magic Kingdom. Skirting fantasy and sci-fi and Western frontiers, Disney's garbage is whisked sightless and soundless through giant underground pipes and tubes to a far-off point on the grounds where it will be baled and carted off, concealed in covered trucks, for its final destination: the incinerator. Out of sight, out of mind. (Estimates are that Disney World generates fifty tons of garbage a day—popcorn boxes, hot dog wrappers, Kleenexes, lost sneakers, broken balloons, and the like.)

Not that garbage disposal is all so earthly and mundane as burn it or bury it. The real far-out action in garbage is from two Atomic Energy Commission hotshots, Bernard Eastlund and William Gough. They call their method the "Fusion Torch" method of garbage disposal or: "Closing the Cycle from Use to Reuse." "The world is now on the steep slope of a rapidly rising transient driven by a population explosion," the two stated in an introduction to their concept. "Over the next 30 years, there appears little probability that the world population can be significantly reduced below the projected 7 billion persons for the year 2000. The result will be extraordinary demands for food and raw material accompanied by growing volumes of wastes and pollutants."

To two AEC scientists their answer lies in the fusion torch, which would use the energy of a hydrogen bomb to vaporize garbage—everything from junk cars and beer cans —down to its basic elements. Their revolutionary idea is based on the fact that sometime within one to five years scientists will have figured out a way to harness the energy of the hydrogen bomb—controlled thermonuclear fusion (as opposed to the atomic bomb's nuclear fission: fission is the splitting of the atom, fusion is the joining together of atomic nuclei to form heavier nuclei, resulting in the release of enormous quantities of energy). When that happens, one result could be the vaporization of garbage, which would serve two purposes, one, get rid of the garbage, two, save mineral resources for reuse. The fusion torch would "close the materials cycle that runs from raw material to people to pollution," says Eastlund, "by forming a bridge between the wastes that are now the end product and the people again. A complete, reusable cycle." As the two stated in their introduction to their preliminary report, "Over the 35-year period from 1965–2000 almost 10 billion tons of solid refuse will have been accumulated in the United States. If compacted and buried to a depth of 20 feet, a land area the size of the State of Rhode Island would be required." As they see it, "the fusion torch concept offers the ability to meet the urgent future needs for raw materials and to eliminate the buildup of wastes."

Professor Robert C. Bostrom (geological sciences) and Dr. Mehmet A. Sherif (civil engineering) from the University of Washington figure that the best thing to do with leftover garbage is bury it in the ocean's trenches. For one thing, these trenches are very deep and getting deeper as their bottoms drive farther into the earth because of the downward wrinkles in the earth's crust that push toward the interior. For another thing, the professors claim that anything dropped in these trenches will, with geologic slowness, fall many miles down into the earth. Once having made this Fantastic Voyage to the Center of the Earth the garbage will stay there, rather than washing back on our shores and beaches. "These trenches are the only places on earth where things go down," says Prof. Bostrom. "No matter where else you put something—in a mine, on the ordinary ocean bottom—eventually it will come back." (The two professors want as much recycling as possible to be done, however. To them the only good garbage is that which has been mined as much as possible for economically recyclable goods. Once that is done, the garbage would be compacted to be heavier than water, then sunk into the trenches.)

Mike Creedman, a free-lance film-maker on the West Coast, was involved in a film project on the power siting problem facing this country. As our voracious demands for electrical power go up and up, more power plants have to be built. But nobody wants a power plant in their back-yard, fossil-fueled or nuclear-powered. Air pollution, de-struction of the landscape, fear of nuclear energy, all these things conspire against the siting of power plants near energy sources or near high-demand areas. Creedman, as he researched the project, hit upon a solution: build ocean islands out of garbage—compacted, made into bricks, turned into landfill—then put the power stations on the islands. He approached a Consolidated Edison executive with the idea. "At least he didn't throw me out of the office." In fact, Creedman figures the executive might even have been intrigued. Even New York City EPA head Jerome Kretchmer is an offshore island buff. "We'll never run out of landfill," he told a community group late in 1971. "We can always build offshore islands with landfill, then put things on them. Parks, housing, power plants. I don't worry about landfill."

Then there is Dr. W. Dexter Bellamy, a biochemist at the General Electric Research and Development Center in Schenectady, New York. Now Dr. Bellamy has really come up with something to dispose of our garbage: bacteria. Very special, scientific bacteria. It eats cellulose. According to one study, up to two-thirds of the garbage deposited in the city dump consists of various forms of cellulose: newspapers, paper products like tissues and packaging, certain foods, such as lettuce and cabbage leaves, and cotton fabrics and lawn leavings. Dr. Bellamy's plan is to have this cellulose-rich garbage eaten by his bacteria and turned into protein-rich animal foods. "In the United States, cellulose and its chemical relatives are discarded at the rate of about three pounds per person per day," said Dr. Bellamy's boss, Dr. Arthur Bueche, GE vice-president for research and development. "Waste cellulose would be converted into an economic asset by Dr. Bellamy's approach." That approach is Dr. Bellamy's one-celled bacteria, which he found while browsing through compost heaps in the Schenectady area, and in refuse from a local paper mill and in dirt around hot springs at Yellowstone National Park. His microbes are thermophilic (heat-loving) things who live at temperatures of 130° to 180° F. A trillion of them would barely fill a tablespoon. So far Dr. Bellamy has isolated strains of bacteria that digest cellulose rapidly, reproduce quickly, and produce what he calls a "biomass" that contains a high percentage of protein. He concedes that much more research and development must still be done, even before a pilot plant can be built. None of the problems are insurmountable, he insists, looking forward to the day when large tanks of thermophilic bacteria abound about the land converting our garbage into animal fodder.

Given our penchant for packaging (51.7 million tons per year) it is little wonder that there are plans afoot to give it special treatment. The United States Department of Agriculture is at this very minute working on an edible food-packaging material. Instead of plastic wrap around your turnip, it is conceivable that in the future you will buy a turnip all cleaned and prepared and wrapped with a packaging material that you simply eat right along with the turnip. No excess packaging to pay for. What the USDA is experimenting on right now is what they call a "relatively stable fat which solidifies to a non-greasy yet flexible solid." According to the USDA researchers, the problem is that so far non-greasy and flexible are mutually exclusive properties. But they are working on it, using such things as peanut oil and cottonseed oil. (Or, as they call it, "modification of cottonseed, peanut, and similar oils by introduction of acetic acid to produce a large production of diacetoglycerides.") If this comes to pass, it opens up a whole new world for eggplants and turnips and carrots and apples.

In the meantime the only known man-made edible packaging is the ice cream cone.

VI

Turning a Sow's Ear into a Silk Purse

We in the United States are the direct descendants of a pack of ragpickers, string savers, foil hoarders, and paper bag collectors. Confront a housewife with a coffee can and she responds with ecstasy and a pencil holder. Think of all those Quaker Oats boxes holding balls of yarn and crochet thread. The day the lady in Nebraska figured out how to make a piggy bank out of a one-gallon plastic bleach bottle was the day American technology should have given in, given up, and gone home.

It is little wonder then that when a new word entered our vocabulary a few seasons back, a few common, ordinary string-saving Americans found little difficulty in responding positively. Recycling was something most of us had, in one form or another, indulged ourselves in everyday. Sometimes it was difficult. How many one-pound coffee cans can a girl save without getting some sort of reputation in the neighborhood? Hence, for a long time we were confronted with nothing less than the closet ecologist: the person who not only hid his light under a bushel, but closed the door behind him. Then came the 1970s, and the world's biggest coming-out party was held. All over the country, ecology freaks stormed out of their closets and joined a long line of other minority groups demanding a fair shake from the rest of society. Homosexuals, Black Panthers, Chicanos, women. The anti-warriors of the pro-America movement.

Recycling caught the spirit and imagination of America. "Either we continue on the same self-defeating, downhill road we have been traveling," said M. J. Mighdoll, executive vice-president of the National Association of Secondary Material Industries (NASMI), "a road filled with dumps, landfills, smoking incinerators, litter—or we can take a new, hard look at these solid wastes. We can see them for what they are: potential new resources, waiting to be recycled and put to productive use for our society."

And so it came to pass that a new word was given unto us: recycling. Take things, use things, return those things to the natural circle of things. Buy a newspaper, read it, recycle it so newsprint can be made of it, so it can be bought, read, recycled, reused. Etc. Buy a can of beer, drink it, recycle it. Recycle those 62 billion cans we use each year. Recycle those 43 billion bottles and jars. Those 57 million tons of paper. Keep the circle closed. Don't break out of it, because to do so means Trouble.

For recycling is indeed a revolution. Earth Day was the great catalyst in our ecological lives. That sunny April day in 1970 woke us up. Stirred the body politic to something besides rhetoric and retorts. Until that moment the ecology movement was a small movement, mostly confined to the sprawling sweep of David Brower's Sierra Club and its fights to save whole forests and whole mountains and whole chunks of the country. Nobody had, so far, brought it all back home to America's backyard.

To some activists—radical and liberal alike—it seemed there were indeed more pressing fights to be fought. We were still miserably embroiled in that miserable little war in Vietnam. Our cities were decaying, our children were starving or dying of lead poisoning or rat bites. Junkies nodded out in our streets, our children tripped out in our living rooms. There were more immediate and more pressing fights, or so it seemed.

But upon examining the problems that the ecology movement is focusing on, we find they are all part and parcel of those other seemingly more pressing problems. A country so wrapped up in growth and expansion it cannot sit back for a moment to determine where it is all taking us. A country whose priorities are so warped it can earmark $300 million for research on biological warfare and withhold an allocated $5 million for research on lead poisoning (a major killer and deformer of ghetto children who nibble away on lead-based paint in their tenement apartments). As our air and water and land became more and more polluted, we began to see how all these problems were interrelated. To talk about poverty is to talk about superconsumerism. To talk about war is to talk about twisted priorities where a world of peace would be so fouled it would hardly be worth living in. To talk about justice for all was to talk about industrial pollution that was all but encouraged by the federal government, which neither imposed tough standards nor saw to it that those existing were even enforced. What good would it do to win all the other battles, when we lose the war by dying of emphysema or mercury poisoning or go crazy because of the noise of trucks and cars and jet planes. Or, worse, are buried under a pile of garbage next to a denuded forest and a stripped mine, naked of natural resources.

Hence, we justified our activities in Earth Day. Recycling became the new password to the good life.

In the words of Mr. Mighdoll of NASMI, "The moving on—to new land, to new forests—is gone. We who have lived in an era of seemingly limitless resources now know better. We are victims of our own concurrent success and failure. It has been the nation's successful and overpowering economic drive—the development of such a huge and sophisticated industrial marketplace—that has created the failure. The failure to manage our environment and balance our resources to that economic success."

One of the most pressing areas for recycling and, because of its sheer bulk one of the most obvious, is paper. In some areas, New York City for example, paper hits as high as 80 to 90 percent of the solid waste. Some 80 percent of the paper used is onetime use; only 20 percent is recycled. During World War II we recycled 43 percent of our paper because of war pressures and rationing. The way we recycle today—20 percent of our paper stock—we are already saving upward of 200 million trees a year. There is a major effort on now, spearheaded by both ecology groups and secondary materials groups such as NASMI, to raise that recycling level to 50 percent. The result, they figure, is a "secondary forestland" of 500 million trees, an area equivalent to all the New England states plus New York, New Jersey, Pennsylvania, and Maryland. The National Academy of Engineering estimates that a minimum of 35 percent of our paper will have to be recycled by 1986 just to keep the U.S. wood resources in balance.

Little by little these small sparks of enthusiasm, set off initially during Earth Day, 1970, began catching hold. All over the country, neighborhood recycling projects began blossoming. Boy Scouts in Nebraska, Girl Scouts in Iowa, Keep Los Angeles Beautiful in California, political clubs in New York City along with artists down in the Village, activists on the West Side. Lola Redford, wife of actor Robert Redford, started a consumer newspaper printed on recycled paper, devoted to spreading the word not only about recycling but about the state of our ecological health in general.

New York City's EPA (Environmental Protection Administration) and the independent Mayor's Council on the Environment began pushing big-city big businesses to investigate using recycled paper in their offices. The city, pushed and prodded by the EPA and the MCE and Mayor John Lindsay himself, began purchasing recycled paper for its offices. The city drew up a list of purchasing specifications that required that all bond paper purchased by it should contain "a minimum of 20% recycled deinked fibres." Even boxes purchased by the city had to qualify. "Corrugated cases shall contain a minimum of 30% recycled fibres (by weight)".

In New York City, Consolidated Edison began experimenting with the use of envelopes made from recycled paper for its monthly billings: that sounds small, but Con Ed uses up 30 million envelopes every three months. Each year the Wells Fargo Bank of San Francisco recovers from its own offices 12 tons of tab cards, 6 tons of white ledger paper, and 17 to 20 tons of mixed paper from their record storage department. These papers are sold to scrap paper dealers for recycling.

The New York Telephone Company each year delivers 44 million pounds of telephone books into that city's garbage pile when it delivers the new phone books. Time was when the telephone company would trade new books for old and send the old ones off to be repulped, de-inked, and turned into new books. They dropped the practice for a variety of reasons, not the least of which was the expense of picking up the old books and the prohibitive cost of shipping them to a recycling mill. Jeff Padnos, of the city's EPA, figured those 44 million pounds of telephone books each year cost the city $800,000 to pick up and put down again. "Would they have made the same decision if the city presented them with an $800,000 bill for sanitation services?" Padnos asks. Indeed.

The pressure was finally felt by the telephone company —pressure both from public agencies like the EPA and from private groups like Consumers Lobby for the Environment, a small neighborhood association in Manhattan's Chelsea section. A pilot project was undertaken in Westchester County in New York to collect old books for recycling into new ones. AT&T hopes 1973 books will be printed on 10 percent recycled paper with 1974 aiming for 50 percent.

The first large-scale municipal newspaper recycling project was undertaken in Madison, Wisconsin. Sponsored by the Paper Stock Conservation Committee of the American Paper Institute and the City of Madison, the project got under way September 9, 1968. The plan was to have Madison householders separate their newspapers, some 170 tons, out of their garbage each week. These were to be bundled, tied, and placed at curbside for pickup. A special hopper was designed for the garbage trucks, a metal rack under the body of each compactor truck. Once they were installed, newspapers were collected, sorted, and hauled to a dealer who had contracted to buy the waste paper from the city. He, in turn, sold it to a nearby paper mill.

First-year statistics, according to Edwin J. Duszynski, director of Public Works in Madison, were depressing. Operating costs were $25,000 over and above the payment received for the paper. But Duszynski was optimistic. Space had been saved at the landfill, paper was not blowing around town, and besides, this was just a pilot project.

In 1970, the collection cost dropped from $27.81 per ton to $9.12. Taxpayers can relate to those figures, as can cost accountants, time-study men, and budget directors. By early 1971, the city was actually *making* a net profit of $2.06 per ton. In the first two-year period some 3,242 tons of old newspapers had been collected in Madison, most of which went right back into the economy as new newsprint once it had been repulped and de-inked. Madison currently recycles about 30 percent of her newspapers, compared to about 20 percent nationwide.

(Madison housewives, on whom most of the burden falls, are credited with much of the success of the project. The housewife is the one who saves the paper, bundles it, and sees to it that it gets to the curbside. Further, if she sees sanitation men tossing it into the compactor rather than underneath to the paper rack, "she will immediately phone a complaint," says the Paper Institute.)

As has been pointed out, technology has put so many synthetics into our environment since World War II that we have broken out of the natural circle of things. After World War II there arrived the coated paper package most generally associated with milk cartons. As early as 1960 Americans were already using 10 billion paper milk cartons annually, each one coated with waxes, plastics, resins, adhesives. Until recently, there was no hope for this particular waste. It was too complex, too expensive, too much trouble to separate all the substances which had been so firmly bonded together.

The Riverside Paper Company of Appleton, Wisconsin, is a paper company that owns no forests and must rely on paper itself as its chief source of paper fibers. The Riverside engineers were able to separate those two substances—the plastic from the paper package. Not only does the process, called "Polysolv," melt off and separate the substances but the jellylike gook that is left over—the wax and plastic and adhesives—is in turn thrown in with ordinary fuel oil and used as fuel to fire the paper mill's steam boilers. Riverside's Polysolv product, which they call "Ecology," uses up 50 tons of coated waste cartons per day, thus saving 850 pulp wood trees every twenty-four hours.

New York harbor used to be one of the busiest ports on the East Coast. No more. A combination of antiquated equipment and stultifying union rules has switched most commercial shipping elsewhere, particularly over to New Jersey. All along the Hudson River rotting piers and decaying docks sullenly sink, little by little, into the river. When they go they create still another environmental nuisance: floating debris. A few enterprising firms have looked at those rotting piers and seen new natural resources. In New Jersey, for example, in the shadow of the Statue of Liberty there is, appropriately enough, the Liberty Lumber Company, which has been salvaging this harbor debris and producing fresh lumber from it. Recycling an average-size pier yields about a thousand piles thirty to sixty feet long. Every pile represents one tree.

To think of recycling metals is to think of World War II and scrap metal drives. Indeed, the most obvious metal to be recycled is the ubiquitous can, both "tin" (food cans) and aluminum (most beer and soft-drink cans). Beyond that is a whole spectrum of scrap metals to be found in cars, aluminum foils, copper wiring, and carpet tacks. Steel, from which our tin cans are made, is already recycled in large quantities: some 52 to 57 percent of the total steel production each year is recycled back into products and, hence, back into the economy. Some 29 million tons in 1970 were recycled. Part of that is simply the industry cleaning up after itself. And part is due in no small part to scrap dealers, or junk men, if you will. Dealers in secondary materials, to be more polite. Indeed, the neighborhood junk man is the original recycling center.

Part of the problem recycling is looked to to solve is garbage—it keeps piling up. Part is the salvation of natural resources. "Even if we solve the garbage problem," Robert F. Testin, director of Environmental Planning at the Reynolds Metals Co., says, "we still have another more basic one—the depletion of natural resources. Iron, aluminum, copper, and tin exist in high concentration in only limited mining sites. Even forests are only semirenewable and cannot indefinitely keep up with continued demand for pulp and paper." Testin's company, Reynolds, began its own in-house reclamation service back in the late 1950s to reclaim all-aluminum motor-oil cans from service stations. In 1967 Reynolds opened the first aluminum recycling center in what they term "high-aluminum-use areas" in Miami and Los Angeles. By 1970 Reynolds had eight more centers around the country, plus a deal with brewers to serve as satellite collection points, some 250 in all, in fourteen states. More than 4 million pounds of aluminum was collected by Reynolds in 1970 alone, and $400,000 paid out for it.

By 1971, recycling centers were going strong with more and more companies, both can makers and users, joining in. There were five hundred locations in twenty-one states. The Aluminum Association figured that the total 1971 aluminum haul would amount to more than 515 million cans—nearly 25 million pounds—and a payment to groups and individuals of nearly $2 million.

It was a start at least. Because old aluminum cans make dandy new aluminum cans. Additionally, private groups such as churches, schools, block associations, ecology organizations, and political clubs around the country set up their own private collection centers to supplement those five hundred industry centers.

The auto is a different story. Particularly those 7 million cars that are discarded in America each year; the cars may be junk, but they also contain large quantities of iron, steel, and aluminum. "The automobile, of all the individual items that we use, probably represents the greatest source of recyclable materials in the United States today," says Karl C. Dean of the U.S. Bureau of Mines office in Salt Lake City. As Dean figures it, the automobile industry consumes 50 percent of all the rubber and lead used in the United States each year, along with 35 percent of the zinc, 20 percent of the steel, 13 percent of the nickel, 10 percent of the aluminum, and 7 percent of the copper.

The U.S. Bureau of Mines estimates that from each one of those 7 million junked cars there are valuable and easily recoverable metals: over 3,000 pounds of iron (worth $35), some 40 pounds of copper (worth $11), 54 pounds of zinc ($3), 50 pounds of aluminum ($6), and 20 pounds of lead ($1.40).

Currently automobile shredding plants, which can chew up a car and spit out fist-sized pellets in a matter of minutes, are located all over the United States, primarily near major metropolitan areas. These automobile shredding plants— usually a hammer-mill/shredding operation—now work over some 3.5 million junked cars per year. That still leaves 3.5 million a year to go.

There is, near Funabashi, Japan, a high-speed auto salvager that has been in operation since 1966 crushing and cooking 50 cars an eight-hour day. Cars are simply sent through a super-assembly line that alternately preheats them and cooks them to burn off nonmetallic materials and melt down the nonferrous metals (everything but iron). The remaining ferrous material is then compacted into a bale with an iron content of about 98 percent, which is sold for scrap. The Japanese were getting about 30 tons of iron scrap a day and figured with the bigger American cars the load would go up to a ton of iron scrap and a half-ton of nonferrous scrap metal.

The Hyman-Michaels Company of Chicago—a company specializing in iron and steel parts and in the dismantling of railroad cars and locomotives—imported a Car-B-Cue, which bakes junked autos, with an eye toward marketing the process around the country. Then the boom was lowered. "We have the only Car-B-Cue installed in this country," says Ralph Michaels, president of Hyman-Michaels. "It is only in partial use. Unfortunately for us the city, county, and state air pollution control boards are continually changing the rules of the game. Every time we spend money to upgrade and control the effluence from the burning chamber we find ourselves faced with a more stringent law." About a year ago Hyman-Michaels made the decision: put the big oven in mothballs. They now operate only the car press, the part that squashes the car down into a mixed bale of iron, nonferrous metals, and glass.

All over the country, new industries are springing up to take care of the dead and dying in our automobile population. While a Car-B-Cue is still a dream, there are practical and operable approaches to the problem. The hammer-mill/shredder process—built by scores of companies around the country—does just what it says it does: hammers the car, shreds it, and then separates out the materials, leaving iron and nonferrous metals for sale as scrap. Pollack-Abrams for example, is a hammer-mill/shredder operating out of Philadelphia, with a capability of processing 600 cars a day. Obviously the biggest market is around cities, where the most cars are junked. Since profits come only on volume, any kind of hammer-mill/shredder (or Car-B-Cue or whatever) has to be around its own very specialized market, for they have to operate at maximum level to be profitable. Already "car wars" are shaping up around the cities as those people into hammer-mill/shredding vie to corner the market to keep themselves supplied with auto bodies.

But let us not neglect glass, which constitutes 6 percent of our 360 million tons of garbage. Before World War II the typical glass bottles made about forty return trips. Take out a bottle of Nehi Orange, drink it down half-warm under the shade of a healthy dutch elm tree, return the bottle for the two cents deposit. That old Nehi bottle would then be sent back to the beverage company, where it would be sterilized and refilled and sent back out into the world again. Then came those pesky, lighter-weight nonreturnables. And the resulting solid-waste problem. "The answer lies, of course, in the fact that the consumers of these products want, need, and demand the convenience of the nonreturnable container," insists Richard L. Cheyney, executive director of the Glass Container Manufacturers Institute (GCMI). "And this demand continues to grow," he says of that industry-initiated program of no deposit–no return bottles. (No housewife dreamed up the no deposit–no return. No clutch of consumers, sitting around trying to figure out ways to make their lives easier in the supermarket, invented the no-return bottle.) "Under our present mode and standard of living, we find that housewife demand has created a substantial market for malt beverages and soft-drink containers that do not have to be returned to the store."

Just in case all those arguments did not hold up, the GCMI got to work and jumped on the recycling bandwagon. And even they had to admit, at least in public and in their press releases, that it was working. First they opened a pilot program in Los Angeles in the spring of 1970. Bring us your tired bottles, they told the residents. The L.A. plant was barraged with bottles—some 1.5 million bottles a week are brought there. Even by the glass industry's pessimistic standards, the program could go nowhere but up. There are now a hundred glass-container redemption centers in twenty-five states. Every year those crazy housewives who won't return a bottle to the store right around the corner will walk, drive, hitchhike, crawl, and otherwise find their way to a redemption center in such force as to create an annual bottle return of some 793 million bottles and jars for which industry pays out $4 million a year.

"The true meaning of the bottle reclamation program is that it is a first step toward the day when it will be possible to mechanically separate all reuseable components of solid waste at municipal or regional collection centers for recycling into primary and secondary products," Cheyney told the annual meeting of the Keep America Beautiful people in 1971. It had been overwhelmingly proved that people are not only willing but exceedingly able—793 million bottles' worth—to return their used glass containers. And all the glass industry can see is that it opens up the possibility of ways to mechanically separate out glass from the rest of the garbage.

The bottle and can manufacturers are convinced, for one reason or another, that the returnable is a dead issue. They are convinced people (1) won't buy them and if they do they (2) won't return them. They are furious and frustrated at all that legislation brought about, they say, by crazed ecology enthusiasts who do not know the first thing about (1) cans and bottles and (2) the American consumer. They argue that bottles would have to be heavier and therefore cost more; they argue that returnables of any sort—bottles or cans—would have to be processed, and this is also costly. Of course, they admit, the cost would be passed on to the consumer. Profits must be high to satisfy the stockholders who are, in turn, consumers. Besides—Cheyney winds up and lets fly with the tear-in-the-eye argument—"banning the one-way container could cost the economy of this country up to $10 billion and dramatically affect about 165,000 jobs."

Thus it would be expensive to process returnables—hire

extra help to clean, sort, etc. And 165,000 might lose their jobs if one-ways were banned, thus cutting down on the number of containers made, etc. *That* set of arguments will take some sorting out itself.

Naturally, industry shies away from discussing the jobs lost when they converted to disposables. Jobs in supermarkets sorting bottles, jobs in bottling plants unloading, sorting, and sterilizing those returnables. Nor do the can and bottle people talk about the amazing decline in the number of breweries, for example. Beer consumption is rising steadily (up 33 percent). But from 1968 and 262 breweries across the United States that figure has fallen to fewer than 80 because of industry consolidation of plants—and the switch to no deposit–no return bottles and cans. According to a U.S. Department of Commerce census, that consolidation alone has cost nearly 20,000 jobs. Soft-drink figures seem to be following the same pattern. Owens-Illinois, one of the biggest manufacturers of soft-drink bottles, has a thousand local bottling plants today; it estimates that in ten years it will have fewer than a hundred plants in the country.

Meanwhile the container industry has used the labor argument over and over to kill legislation restricting the sale of disposables. Concurrently, the possibility of creating jobs in a revitalized returnable market is, of course, solidly rejected.

Cans, too, were being returned at an alarming rate to reclamation centers—some 95 million at industry collection sites alone—for a total of 600 million. Indeed, the whole industry-sponsored recycling program almost overwhelmed itself as bottles and cans came pouring into recycling centers.

The 793 million returns of glass containers were made at the industry's own reclamation centers—bottling plants and factories—at a hundred locations in twenty-five states, which is not overwhelming the public with chances to return. Notwithstanding, the bottles came crashing in. Nor does that enormous figure of 793 million take into account the hundreds and thousands of·citizen recycling centers that began springing up around the country in 1971: Boy Scouts and block associations and supermarkets and church basements and social clubs. When those groups are added in, the figure of items returned for recycling would skyrocket.

As for returnables, even with the downgrading by industry the average returnable bottle still makes nineteen round trips. (In cities, however, that drops alarmingly to four circuits.) That seems to say something. Returnables, by the way, account for about 10 percent of the total number of glass food containers used; about 2.7 billion out of a total of 29.4 billion units.

"My dream is a standard bottle that everything—wine, milk, gasoline—could come in," says Jerome Kretchmer, NYC's flamboyant EPA chief. "All in the same quart, half-gallon, and gallon jugs. The business of manufacturing the bottles could be replaced by the business of collecting and washing the bottles with high-energy steam."

Bringing up the rear in the fight to recycle our way to victory, is, of course, plastic. "The plastics industry has not been able to take part in the movement towards redemption centers," the Society of the Plastics Industry (SPI) said in a statement before the New York City Environmental Protection Administration in 1971. "The reason for this is that we do not have adequately developed techniques to recycle or re-use plastics from municipal refuse." If forced, either by public opinion or public law, to do so, "we would have to turn them over to the Sanitation Department for disposal. Currently, there is no real economic value for used plastic containers." There is the rub: no real *economic*

value. No matter that a towering mountain of costly garbage—part of it plastic—is threatening to fall over on us.
No matter that consumer convenience—at a cost of millions—is warping not only our priorities but our garbage
cans.

Recycling is not so simple as bundling your newspapers
and stomping your tin cans. First of all, there is public
apathy: some folks either cannot see the garbage crisis or
cannot be bothered with garbage the way they cannot be
bothered with elections and Southeast Asia and Appalachia
and women's rights. Besides, garbage has no obvious glamour, no sex appeal.

But more important are four obstacles built into our
whole socioeconomic-political structure that make the
whole concept of recycling both less than practical and
more than troublesome.

A series of discriminatory freight rates, for example,
makes the shipping of virgin materials—pulpwood trees,
mineral ores—much cheaper than used materials such as
bundled newspapers, stomped tin cans, and broken glass
from collection centers. To ship virgin timber from Bur-
lington, North Carolina, to Roanoake Rapids (a distance of
120 miles) costs $1.99 per ton. To ship paper waste between
the same two points costs $3.60 per ton. From New York
to Chicago virgin pulp is shipped at twenty three cents per
hundredweight. The rate for paper waste is forty cents per
hundredweight. It's the same story in metals. To ship raw
ores from Boulder, Colorado, to El Paso, Texas, for example, costs $14.07 per ton. To ship scrap metals the same way
between the same two points costs $17.60 per ton.

Federal tax policies are the black lung disease of the
ecology movement in this country. Over the years of this
country's childhood, Congress passed laws to encourage
the development of our own natural resources, which for

decades did indeed seem endless. Today, we know they can be depleted. The depletion allowances and capital gain treatment for natural resources continues. Companies were and still are given a big go ahead to deplete our natural resources and a tax break to boot.

Politically, in addition to the federal tax policies, the federal government for years worked at cross-purposes with the environmental interests. Many government specifications, particularly for paper, called for virgin materials. Only recently has the federal General Services Administration begun developing new specifications that allow the use of products made of recycled materials.

It is significant to note that all of the recycling centers set up after that first magnificent flush of conscience and consciousness inspired by Earth Day were strictly volunteer. They were either citizen-sponsored or, in part, industry-sponsored. No city, state, or federal government made either possible or easy. These ecological stirrings were strictly grass roots, strictly volunteer. If cans had to be stomped, it was a volunteer who took time off from baby-sitting chores or classes or work to do it. If papers had to be bundled or stacked—ditto. No separation laws were written and enforced; no municipal collection centers were set up. In short: the people decreed recycling but the governments of the people did not follow through on that mandate.

Last, but certainly not least, has been public attitude toward recycled products. In the language of the paper trade, for example, paper made from recycled materials has been referred to as "bogus." The general use of the terms "virgin" vs. "secondary" materials says it all. And face it—there's something about doing it to a virgin. Felling a forest of tall, proud-standing trees does a lot more for the ego than dealing with bundles of newspapers tied up by some housewife out in Madison, Wisconsin.

Essentially, then, recycling to provide a new use for much of the garbage in this country is a two-sided problem. It is one of collection and it is one of finding markets for what has been collected. Paper is the easiest to handle at the local recycling center. Until recently, however, it was the most difficult to find new markets for. Both the metals and the glass industries were able to absorb those cans and bottles. Paper, however, had the additional problem of a social stigma. Big firms—who feel their very status is at stake in the paper they write letters on and the enevelopes they send their bills out in—objected both violently and firmly to the idea of using recycled paper in their offices. For example, William Cleveland, purchasing manager of Goldman Sachs, one of the bigger Wall Street brokerage houses, flatly stated late in 1971 that his company had nothing to do with recycled office papers because they had their "image" to protect. Other factors involved in switching to recycled office papers were cost and availability. These, for a time, were much more meaningful and cogent arguments. When the raw material (i.e., old paper) for recycled paper was not so readily available and when the market for recycled paper products was small, the cost was up and availability was down. As the whole idea of recycling began catching on here and there all around the country, the raw materials became more plentiful, markets for recycled paper products expanded, and, naturally, the price went down. Those new GSA specifications did not hurt, nor did New York City's insistence on recycled paper. Large companies (Canada Dry, Chase Manhattan Bank, New York Telephone, Bank of America) gradually switching over to various uses of recycled paper—and garnering a lot of free publicity for doing so—helped. Newspapers (the Baltimore *Sun*, *The Washington Post*, the Philadelphia *Inquirer*, the New York *Daily News*) even started printing on newsprint made from recycled paper.

The whole environmental movement is the offshoot of many long years of patient work by the conservationists in this country, who each day were able to convince a few more people that things really are interrelated. That the whole world is caught within a fragile circle, and to break any link of that circle is to change the course of natural history. A few visionary souls got Earth Day, 1970, off the ground and we are where we are today because of it: in the forefront of America's new revolution. Few people either knew or cared about the environmental crisis facing this country prior to 1970. They were either somebody else's problems or problems so prevalent as to be normal and commonplace. Few people saw them as interrelated problems coming from technology, affluence, consumerism, population. Nobody saw them as problems interacting on our very existence or the quality of our lives. Now, at least —somebody cares.

VII

How to Experience Joy Through Stomping Tin Cans

Down in Missouri, buried back in the Ozarks amidst Minnie Pearl's Fried Chicken Parlors and Lake o' the Ozarks chenille bedspread factories, is a laboratory fairly bursting with activity. For it was here that Dr. Ward Malisch, a mild-mannered young man from the University of Missouri School of Mines in Rolla, began working out his plan to save America from herself. Dr. Malisch and his co-workers, Delbert Day and Bobby Wixon, were huddled in their laboratory figuring out a way to cope with glass. They toiled and tested, worked and worried, and finally they had it. And they called it Glasphalt. For, indeed, they had come upon what might be the most perfect—if not the most ironic—final resting place for all those beer bottles and soda-pop bottles and wine bottles that' lurk on darkened highways and dim curbsides just waiting to puncture your new Tiger Paw. Their idea: grind it up and make paving material out of it.

Today, success is, if not right around the corner, at least underfoot. The U. of Mo. has already paved the parking lots at the Rolla campus with patches of Glasphalt along with a 20- by 600-foot test strip on the campus. Skid tests have been conducted and the results so far are good.

On the east coast the Bureau of Sports Fisheries is developing breeding environments for marine life by heaving old tires onto the ocean floor. They settle there, and various and sundry forms of fish gather to, in the words of Cole Porter, do it. The State of Florida is experimenting, quite successfully, with using tires as breeding grounds for oyster farming where oysters, ditto, do it.

Additionally, old tire rubber is being considered as a component in blacktop dressing for driveways and parking lots. The Department of Health, Education, and Welfare (of which the Environmental Protection Agency and the Bureau of Solid Waste Management are a part) awarded just such a research contract in 1971 to the Battelle Columbus Laboratories. Benson G. Brand, project director, figures that old rubber tires in blacktop surfacing will increase durability, resiliency, and resistance to abrasion. In 1971 the Goodyear Rubber Company, in a project shared with the Cities Service Company, began experimenting with using ground-up scrap tires to produce carbon black, one of the key ingredients in tires. (Normally carbon black is produced commercially by burning raw oil under conditions of incomplete combustion.) Under the Goodyear–Cities Service plan ground-up tires are mixed with the oil before combustion. According to Goodyear, the rubber from one passenger tire provides enough carbon black to produce a new passenger tire. "There is an outstanding example of recycling," says a Goodyear spokesman, "a continuous circle involving the production of new tires, the total destruction of old tires, and use of the materials in new tire production." Goodyear figures this process could use up about 60 million scrap passenger tires a year.

At Texas A. & M. research engineer Douglas Bynum is working on using discarded rubber tires to give asphalt more flexibility and more resistance to cracking. Working in the university's Transportation Institute, Bynum conducted tests using asphalt plus ground-up scrap tires. His test results showed that the powdered rubber—used as a binding material—increased asphalt's overall cohesiveness so that it does not split when roadbeds shift slightly or sink. "We could save the money we spend to dispose of these materials and get better highways," says Bynum. The way he figures it, old tires and bottles in combination with asphalt could pave a freeway that would span the United States twenty-three times.

Down the road a piece in Lubbock, biologists at Texas Tech. are working on a process to convert various kinds of garbage into a protein-rich product for use in meat production and, possibly, for human use. What the Texas Tech. scientists are doing is much like Dr. Bellamy's work with bacteria at the G.E. laboratories: breaking down things rich in cellulose—newsprint, waste paper, weeds, feedlot waste, and a collection of other throw-aways—by using what they call "carefully selected bacteria" to digest the cellulose content of this and turn it into protein.

These "carefully selected bacteria"—bacteria weeded out from the hundreds who apply—are turned loose on a batch of garbage fairly reeking with natural cellulose. After the protein bacteria get through the newsprint and mesquite and cotton ginnings, the high-protein residue the scientists end up with is a powdery substance looking like a cross between sugar and heroin. Protein is vital to human growth. And we don't get all we need from our corn flakes and enriched white bread. A 1,000-pound steer produces only 1 pound of protein per day. But 1,000 pounds of the Texas single-cell protein bacteria can produce 100 trillion pounds of protein per day. Just on a steady diet of newspaper and cow manure and the like.

The U.S. Bureau of Mines is working on a method to make lightweight building blocks by molding concrete around cores of compressed automobile scrap. Car bodies are burned to get rid of flammable materials—plastics, seat stuffing, wood paneling—then the scrap metal residue is cut into sections and compressed into cubes, which are each encased in two inches of concrete. The heat generated during the incineration process is used to steam-cure the concrete coat.

Dr. Altheus Spilhaus, scientist and former dean of the University of Minnesota Institute of Technology, has an even better idea for disposal of discarded cars. The way Dr. Spilhaus sees it, every year thousands of carloads of prime Kansas beef are shipped east to New York City, where they end up on the menu as prime sirloin and New York cuts and filet mignons. As for the boxcars, Spilhaus contends they return to Kansas empty. His idea is to fill them with discarded autos for shipment back to Kansas, where, he says, they could be used to build mountains and thus break the monotony of the Kansas plains that stretch on and on, an infinity of wide open spaces.

Since 1970 the U.S. Bureau of Mines has been perfecting a system of separating out basic components from incinerated garbage. The plant it has built employs existing technology: magnetic separation, screening, grinding, shredding, burning. What has been added is sophistication.

From a ton of municipal garbage, the device can separate out 700 pounds of iron, 40 pounds of nonferrous metals (including aluminum, zinc, copper, lead, tin, and small amounts of *silver*) and 1,000 pounds of glass. The device is so sophisticated it can even separate glass into its various color categories—clear, green, brown.

From the various kinds of glass the bureau is able to separate out of its garbage, it is making building bricks by mixing up a batch of 70 percent glass and 30 percent clay. The bricks "meet or exceeded specifications for severe weathering," according to the bureau. Additionally, it has discovered still another use for its residue glass: mineral wool, used in building insulation.

The bureau's Coal Research Center out in Pittsburgh has discovered a way of turning ordinary old household garbage into crude petroleum. By combining it with carbon monoxide and steam it is able to get a heavy oil with a low sulfur content. The bureau estimates that each ton of dry refuse would yield over two barrels of this kind of petroleum. "Based on current generated domestic refuse and animal manures, this represents a potential equivalent of 2 billion barrels of oil annually," figures Charles B. Kenahan of the Bureau of Mines.

Another group of scientists is working with garbage at the Illinois Institute of Technology in Chicago. Led by S. A. Bortz, these people, part of the Ceramics Research Division of IIT, have been working on incinerator residue. To most people the junk left over after garbage has been burned looks just like that: junk. To IIT, however, it looks quite different. The scientists take ordinary household garbage—beer cans, potato peelings, coat hangers, aluminum foil, soup cans, tomato-juice jars, soda bottles—and burn it up at a very high heat, 2800° F. as opposed to the usual 2000° F. of most municipal incinerators. The organic materials are destroyed in the conflagration—the egg shells, coffee

grounds, paper packaging, lettuce leaves. Then the researchers take the residue and work with it. It contains,
though hardly in recognizable form, glass, nonferrous alloys, and ferrous alloys. They look upon this residue as
"urban ore," a new kind of natural resource for manufactured products, things like cast doorknobs and round construction pipes, sewer pipes and structural building blocks
looking very much like those cinder blocks used around the
country. Since the sewer pipe is a high-volume material—
approaching 2 million tons annually—it is an excellent application of the urban ore glass, Bortz points out.

As for just plain litter—that miscellaneous flotsam and
jetsam of our affluent lives—there is even a plan afoot to use
it underfoot in road-building. Nearly a hundred acres of
roads and parking lots at Dulles International Airport,
near Washington, D.C., are now paved with a combination
of reclaimed rubbish, bottles, garbage, lime, and water, all
of which was mixed with two leftovers of industrial waste:
calcium sulfate and fly ash. Calcium sulfate is a by-product
of sulfuric acid used in steel processing, and fly ash is that
junk from incinerators that floats down and lodges in your
eyes. As the government put it, in their overstuffed jargon,
this is a "significant breakthrough" in both engineering
and garbage disposal.

One of the more unique uses developed for garbage,
however, came in 1968 during the disturbances that followed the assassination of Martin Luther King. Brooklyn's
Bedford-Stuyvesant area—a community that had carried
the level of urban decay to dazzling heights—learned the
tactic of dumping over garbage cans and setting their contents ablaze. There they were, like signal fires from some
long ago Roman legion, burning briskly on the streets of
Brooklyn. Everybody came out and pitched in. Donations
were taken through the night. Everything in sight went—
garbage cans, litter baskets, vacant lots full of airmail

deposits from top-floor tenants. The whole works. (Vacant lots are a particularly good source of urban ammunition. The city of New York, respecting the American Constitution and its stand on private property, makes it, if not illegal, damned near impossible for city sanitation men to enter vacant lots and clean them up. The debris collects, layers and layers of it, over the years, until finally it explodes or is set fire to. Or, ironically, some citizen group will mobilize the better instincts of the neighborhood to pitch in and liberate that litter. Bed-Stuy, and other neighborhoods that quickly learned this particular guerrilla garbage tactic, merely liberated their own neighborhood landfill for their own purposes. The tactic spread, and for a while one would have thought New York City was going up, in smoke and flames, for grabs. Harlem, East Harlem, the Lower East Side. The ghetto areas were united for once, around their burning garbage. Sanitation men, like firemen in a riot, were afraid to enter the burning streets. The city, to head off a full-scale garbage war, removed most of the litter baskets from the corners of trouble spots "for the duration of the troubles." Years later, it was charged the litter baskets still had not made their way back to the corners of Bed-Stuy and other ghetto neighborhoods around the city.

In 1969 the Puerto Ricans followed the lead of their Black brothers. In Spanish Harlem, on Manhattan's Upper East Side, the Young Lords sparked a garbage-throwing melee near 110th Street and Park Avenue to protest the lack of garbage collection services. "It launched us as a group," recalled Yoruba Guzman, one of the leaders of the Young Lords. "At the time it was used as a strategy, an overall campaign issue to mobilize the community." As Guzman saw it, "We realized there is garbage in the streets. Whenever we'd rap with our people on the stoops we'd bring it up. Everybody up here lives with garbage. We could all relate to it." Throwing it was cheaper and better than analysis. Besides, it led to instant gratification. "People liked it—they got rid of a lot of frustrations by throwing garbage." Additionally, it is a constant weapon. "If anything happens this summer," Guzman said the next year, "it's there to use." If the Lords weren't throwing garbage, they would spread it out as a barricade to stop cars.

One of the best old-time uses for garbage—before we started throwing too many Chux and Baggies and other such nonbiodegradable synthetics into it—was compost. Families would just throw the bones to the dogs, the leftovers into the soup pot, and the rest of it—egg shells, coffee grounds, the scrapings from the bottom of the soup pot— onto the garden. With the introduction of commercial fertilizers plus synthetics and the increasing amounts of paper in our society, composting became impractical and old-fashioned. There were a few token attempts to start up new composting operations after World War II, but they all failed. Of the eighteen or so that started up, eighteen or so failed.

There was some surprise, then, to hear that a Hungarian-born Ph.D. had opened yet another compost plant, Ecology, Inc., in Brooklyn, New York. It is, however, a composting plant with a difference. For one thing, there is Dr. Stephen Varro, a diminutive and dashing émigré who once served with the French Foreign Legion and who strolls around his plant wearing a hard hat, tweeds, and an ascot. For another, Ecology, Inc.'s whole approach to composting is different. "We use everything," he says. Where the other companies spent endless hours pulling all the paper out of their raw materials and then digging holes to bury it in, Varro's compost plant leaves it all in there. He reasons it's cellulose and loaded with nutrients. About the only things that are not used are ferrous metals, which are separated out along the line by a magnet and sold to a scrap dealer. Even plastic is used. Ground up in small pellets it becomes a soil aerator like sand or gravel.

Ecology, Inc.—which environmentalists around the country are looking at with renewed interest—is a nondescript cinder-block building stuck amidst a decaying warehouse section out in Brooklyn. The only thing that distinguishes it from its neighbors is that it is newer and, so far, cleaner. Inside, it is a Rube Goldberg contraption in action. Ecology, Inc. is paid $10 a ton to take 150 tons of garbage a day from the city of New York and process it into compost, which is bagged and sold commercially on the East Coast. Garbage trucks drive up to dump the garbage out, and from there it starts on its fantastic voyage that winds up at the end of the line in a bag of fertilizer called "Ecology." In between are mashers and grinders and conveyor belts to move the pulverized garbage along to a series of enclosed trays. There is is heated up both by the natural processes of organic decay and by scientifically controlled temperatures hitting upwards of 148° F. All pathogens (poisons) are killed off, and the remaining material is fully broken down.

The final product looks very much like the linty contents of a vacuum cleaner bag and to this fuzzy-looking stuff Ecology, Inc. adds nitrogen, phosphorus, and potassium to jazz it up. It is certainly not organic fertilizer, but then again it's not garbage.

Whether the Varro approach to garbage is the answer remains to be seen. After all, there are just so many lawns and golf courses and parkways that can take the fertilizer from all the garbage this country produces. And as Michael Hirsch of the New York City EPA puts it, "Varro will have to prove to us he can sell all the fertilizer he produces. He's only in business as long as he can sell it." Which is certainly the all-American approach to the problem in general and garbage in specific.

Another practical application of the garbage produced in

this country is using it as a fuel for heat and/or energy. This is really not a new notion. Paris is lit, in part, by burning its garbage. But the idea has never been explored in this country on any large commercial scale or even with any seriousness as a way to (1) get rid of garbage, and (2) provide a side benefit in the form of heat and/or electrical energy. In typical fashion, we have preferred to blunder through life using virgin materials—roaring waterfalls and strip-mined coal—to produce our heat and electric power and air pollution.

One trash power plant in this country is down at the Navy Public Works Center in Norfolk, Virginia. The Navy, faced with its own garbage, had a landfill (an open dump) that was rapidly filling up. The city of Norfolk was getting sticky about opening up still another dump for the Navy.

In 1967 the Navy's "Salvage Fuel Boiler Plant" went into operation. It is the largest in the world, featuring two boilers, capable of burning a total of 360 tons of garbage per day while producing 50,000 pounds of steam *per hour*. (As Rear Admiral H. N. Wallin pointed out about the project, this is very important because a "typical Navy destroyer sitting alongside a pier requires about 5,000 pounds of steam per hour when she goes 'cold iron' [is immobile, using steam only to generate electricity]." The Navy figures it is saving about $47,000 a year in fuel costs alone in addition to having eliminated an air pollution problem encountered by its old open-dump burning methods.

The only city to seriously consider burning its garbage for anything more than air pollution is St. Louis.[1] The city

---

1. New York City began burning its garbage for electric power as early as 1903 when an incinerator at West fifty-seventh Street provided lighting for a stable and the nearby dock area. Another incinerator in East New York, Brooklyn, provided electricity for neighborhood shops and stores. On November 30, 1905, the city opened a plant on Delancey Street that burned 1,050 cubic yards (then one-fifth

of St. Louis, in cooperation with the Union Electric Company, is currently experimenting with the burning of St. Louis garbage in existing power plants to provide electrical power. The project, designed by Horner and Shifrin, currently burns 250 tons of St. Louis garbage per day, on a twenty-four hour basis, to be supplemented by what they normally use alone: coal. The garbage is prepared by sending it through a hammer-mill process where it is ground into pellets. Then magnetic separation sorts out the retrievable metals. The residue—pellets about one and a half inches in diameter—is then burned in Union Electric's boilers. Enthusiasts of the program figure that sometime in 1973 the Union Electric Company will have the potential capability of consuming over twice as much refuse as is generated in the entire St. Louis metropolitan area of 2,500,000 people. New York City is investigating this method of burning garbage for fuel as a way of getting rid of its massive mountain of garbage.

Plastic, however, remains the biggest problem. At the Illinois Institute of Technology Kurt Gutfreund, a senior chemist in polymer research, has been trying to figure out ways to make plastic containers break down, to impose biodegradability on them, as it were. Although he found it possible to interrupt the molecular structure of plastics he was working with, they didn't quite measure up to accepted manufacturing standards when he got through with them. Or, as he put it, "the enhancement of polymer degradation runs counter to accepted manufacturing standards in that minimum performance characteristics for the product may have to be accepted to accommodate the needed degree of degradability."

---

of the Manhattan/Bronx total garbage load) to light up the Williamsburg Bridge. By 1907, however, studies were showing that it was costing more, because of increasing labor costs, and it would be cheaper to buy from the New York Edison Company.

Despite the defensive attitudes of the plastics industry as a whole, it is being pressured, little by little, to break out of its plastic bags and enter the real world. Inspired by a pinch of this and a dash of that—Earth Day, consumer groups, bad public relations, constant challenges from both environmentalists and lawmakers—the plastics industry is following the lead of the other problem-makers: it is figuring out new uses for its own particular brand of garbage.

"Let there be no misunderstanding about this," insisted Ralph Harding, executive vice-president of the Society of the Plastics Industry (SPI), at the annual meeting of Keep America Beautiful in 1971. "The volume of plastics being recycled from solid waste is very small and still largely experimental."

In Southern California a few dairies are accepting back their empty polyethylene milk bottles. These have then been ground up, cleaned, and used in the manufacture of soil pipe and flower pots. Out in Ohio a Boy Scout troop collected used plastic bottles to be utilized in experiments for making soil pipe. A municipal footbridge is being built in Elgin, Illinois, made of some 25,000 ground-up plastic bottles used as a replacement for sand in concrete. According to Harding, "Tests thus far show that concrete containing plastics in place of 30% of the sand is lighter, crack-resistant, and easier to work."

At the National Plastics Exposition held in 1971 a German machine, the "Remaker," was demonstrated. According to Harding the machine was built especially to use scrap plastics, mixed or unmixed. "This machine can make useful products such as wheels for toys, shoe soles, decorative products, or bird feeders" out of used or discarded plastic. "Our biggest obstacle," said Harding, "continues to be economics: virgin materials are inexpensive and very reliable, while recycled plastics are not that much cheaper."

But where there's garbage there's hope. And just when our garbage was looking its grimmest, there came, rising over our landfill, a bright and shining sun: the Japanese announced a revolutionary new approach to household garbage. Their plan: collect it, compact it, bale it, encase it in concrete, and use it for building blocks. Garbage men went wild. The press went wild. At last: a fitting resting place for our garbage. Technology will out. We will build high-rise monuments both to and with our garbage. "Few things short of war received as much television, radio, or newsprint publicity," said S. Myron Tadarian, San Francisco public works director, of the Japanese idea. Indeed, the world went crazy. Hordes of Americans streamed into Japan to view the pilot plant. Miles of media copy came spewing back. Joy and jubilation poured out of the landfill of the rising sun over these "garbage blocks."

Then came disaster.

Every now and then a devil's advocate would be heard: there was no working plant in operation. A few foolish garbage men—at the mercy of translators—merely took the Japanese at their secondhand word. It was a class hype by the inscrutable Orientals. And nobody got scruted worse than the gullible American press.

Indeed there was no working plant, just a little pilot operation merrily squashing garbage, coating it with asphalt, and—sticking it in landfill. "Outside the Hamamatsu facility there were several hundred compressed refuse blocks awaiting pick-up," noted an article in *Solid Wastes Management* magazine. And not only were they awaiting pick-up, but they were "emitting a strong odor that permeated the air." Which is a very polite trade-journal way of saying they stank to high heaven. Additionally, "the blocks were surrounded by flies attracted by the smell and the refuse in the cubes."

As if all this were not shattering enough, to those garbage-mongers who were busily building castles in the air out of Japanese building blocks there came still another blow. "The whole idea is full of shit," noted Jerome Kretchmer in 1971, back from a visit to Japan. "The fucking things blow up." Gas collects and, according to what the Japanese call "slanderous rumors," the damn things do indeed blow up. Leonard Stefanelli, president of the Sunset Scavenger Company—the private company responsible for picking up and delivering San Francisco's garbage—also visited the plant and noted that "those blocks certainly were not 'odor-free' or 'uniform' as publicized." On the contrary, "they were not even square or cube-shape. They could best be described as foul-smelling, fly-ridden, un-uniform chicken-wire-and-tar blobs." So much for that.

But for the most part these methods, interesting though they may be, are the show-and-tell approach to garbage. To institute any of these methods—road-building material or cellulose-nibbling bacteria or petroleum-producing tires—would take forever, given American priorities and the state of the American lobby. Can you imagine the protective reaction from the petroleum industry when confronted with any large-scale attack on their interests by the scrap-tire industry who could get petroleum out of old Tiger Paws? With the possible exception of energy producing, these methods are, in the truest sense of the word, Ralph Nader's "hospital approach" to our garbage: wait till the patient is in the hospital before you treat him.

What has to be done, it seems, is to rethink our entire garbage philosophy. Figure out how to cut down our garbage production in the first place. Close the circle by recycling and reusing as much as possible. Then, when that is done, apply as many of these new methods of disposal and use as is necessary to rid us of our leftovers.

Or, as one radical environmentalist said allegorically,

"Those bottle cutters are crap. That's the chichi *House and Garden* approach to garbage. How many nonreturnable bottles can you cut up, for God's sake? That's like being a fat slob to justify going on a diet."

You must admit, it is one way of looking at the problem. Beyond that, perhaps the only consolation is in knowing that somebody cares. Somebody out there—the U. of Mo. and Texas Tech. and the Bureau of Mines—likes our garbage and is working both with it and for it. But look at it from the garbage point of view: for a throwaway bottle, to be a part of a Glasphalt parking lot is still a far, far better thing than it has ever done before.

VIII

The Heroics of Garbage

New York City is known for its dedication. Indeed, dedication to one cause or another runs rampant in New York City. This city fights to lead the world in so many areas that to name them would seem an aggressive act of braggadocio. New York City has more Chinese restaurants than Peking. It is rumored that there are more Puerto Ricans living in New York City than in Puerto Rico itself. Beyond that, New York excels in crime, punishment, graft, corruption, carbon monoxide levels, prostitution, poverty, Beautiful People, homosexuality, venereal disease, and Catholic schools.

It is little wonder, then, that the world looks to New York for salvation. It is not hope misplaced. Indeed, New York City is recognized in the best of circles as the garbage capital of the world.

New York City produces more garbage than any other city in the world. (In fact, New York produces more than most *countries*.) In the 1971–72 garbage season—a socko season if ever there was one—New Yorkers finally got their garbage up to 29,000+ tons a day. That is up 2,000 tons from the 1969–70 season. (It seems hard to believe that just ten years before—the 1960–61 season—New York was getting up only 19,463 tons of garbage a day.)

In New York citizens would seem to work endlessly and tirelessly around the clock to maintain her garbage lead. In so doing, the hardworking citizens of New York have created for themselves a mini-mound of trash within the already impressive mountain of refuse our country is currently stockpiling. This only goes to show what a truly dedicated populace can do when united behind a cause. Not only did New Yorkers increase their production by 40 percent in the last ten years, it is estimated that in the next ten years they will push their production up another 63 percent. Another giant leap forward for mankind.

In 1969 alone, more than 500 million one-way bottles were discarded in New York, up 20 percent in two years. Each square mile in New York produces 375,000 pounds of garbage a day. Just to show that it is indeed quality garbage that is bursting forth out of New York City, New Yorkers tossed out 442,000 tons of Sunday newspapers[1] (the *Times* and the *News*) in 1971. In New York City on any given day waste paper in general—from all sources: newspapers, letters, junk mail, paper towels, theater tickets—equals 100,000 trees. Reynolds Aluminum estimates that although aluminum makes up only 1 percent of New York City's total garbage load, a total of 200 *tons* is discarded every day. If all of that aluminum were converted to Reynolds Wrap it would make a roll over 7,500 miles long, roughly the

---

1. Which cost $13.2 million to get rid of.

distance from New York to Cairo. Never are New Yorkers without their garbage. Never are New Yorkers without their army of sanitmen—some 11,000 of them armed to the teeth with brooms, trucks, safety shoes, and union cards— ready to do battle with their garbage, day and night. The enemy never rests.

The rolling stock of the New York City Department of Sanitation numbers 2,010 vehicles. (That is exactly the population of my home town, Albion, Nebraska.) The Department of Sanitation alone—that Army in garbage green —totals 15,131 people.[2]

But this is nothing, mere preliminaries to the main event. If New York produces the most garbage of any city in the world, it only stands to reason that it costs a lot to pick it up and get rid of it. Indeed it does: it costs $113,783,310 just to pick it up ($30.05 a ton). It costs another $27,685,657 ($4.50 a ton) to put it down in incinerators and/or landfills. When everything gets added up—collection, disposal, street cleaning, litter basket emptying, etc.—New York City's garbage bill is a gargantuan $176,246,604 a year.

There are some who do not appreciate what all this means. Considering the fact that the nationwide packaging garbage is anywhere from a low of 13 percent to a high of 20 percent of the total garbage in this country, New York wins the packaging production race hands down. According to New York's Kretchmer, packaging materials amount to 40 percent of New York City's garbage every year. That counts out to 2,350,000 tons of packaging. It costs $54 million dollars a year to pick it up and put it down.

Not that packaging waste is the only thing New Yorkers excel in producing. Take that Sunday *New York Times* for example (which costs fifty cents to buy and ten additional cents to dispose of). The total editions of the Sunday *Times*

---

2. There are also 20,864 litter baskets in New York City.

sold in an average week weigh more than 7 million pounds. When these 7 million pounds of Sunday *Times* are incinerated they donate 87,500 pounds of particulate matter into the air, plus leaving behind 1.75 million pounds of incinerator residue (ashes and unburned portions of the Book and Magazine sections). Kretchmer figures it requires 1,150 man-days just to collect those 7 million pounds of the Sunday *Times*. "This obviously raises the question of whether the taxpayers at large, many of whom, of course, do not read or buy the *Times*, should bear the heavy collection costs," Kretchmer figures, "or whether the *Times* has a responsibility to move into recycling." Figuring those 7 million pounds to equal 3,500 tons of *Times* and further figuring that it costs $34.55 to collect and dispose of a ton of garbage, the Sunday *New York Times* alone costs New Yorkers $120,925 a week just to pick up and put down, or $6,288,100 per year. It's enough to make a person switch to CBS.

(Convincing *The New York Times* to get into the recycling business will be difficult. They own their own forests.)

Besides packaging (2.4 million tons a year) and the Sunday *Times* (182,000 tons a year), New York City also has the biggest and best abandoned-car problem in the entire world. No other city, state, or country abandons automobiles the way New Yorkers do. (Although it is felt in some quarters that many cars are brought into New York by outside auto agitators and abandoned on the city's streets.) Every year some 73,000 automobiles are left to rust in pieces on the streets of New York City. The problem got so bad, in fact, that New York City finally contracted with private firms to come in and haul these abandoned autos off to the junk yards. Prior to this move, it had been the responsibility of the Department of Sanitation.

The day after Christmas is considered by the Department of Sanitation as "one of the biggest garbage days in the year." That is a little like saying the Grand Canyon is deep. All those boxes and ribbons and cards and turkey bones and eggnog cartons add an impressive 4,000 tons of garbage to New York's already impressive 29,000+-ton total. To cope with it, 5,000 sanitation men have to go on double time. (Not that the DOS is the only place in the city to feel the brunt of Christmas: think of all those janitors and supers and landlords who have to struggle with all those garbage cans holding those extra 4,000 tons of garbage.)

Obviously, once it's picked up, it has to be put down. And, despite its leadership in the field, New York is just like every town in America: it burns it or it buries it. Or, as Dick Napoli of the DOS puts it, "No matter which way you cut the cookie—you burn it or bury it."

From the outside the Brooklyn Incinerator is as normal and ordinary as, say, your average maximum-security prison. It is a yellow-brick building with high impersonal and impenetrable walls and few windows. Instead of turrets with armed guards, the incinerator has two big smokestacks—one armed with an electrostatic precipitator—reaching 200 feet into the air. It's the biggest in the world. Garbage trucks parade up to the incinerator steadily during the day, dumping their garbage into a huge pit that itself holds 12,500 tons of garbage. A crane moves back and forth over this pit, like a nervous bird over a cornfield, periodically swooping down to snatch up a load of garbage in its jaws. It can open up and come up with an entire ton in one bite. The crane drops the garbage into bins that feed it onto conveyor belts that in turn feed it into the incinerator ovens.

Those ovens are a combination of Dachau and the Betty Crocker test kitchens. Four ovens, 30 by 7 by 200 feet, each capable of burning up 10 tons an hour at heats averaging 1600° to 1800° F. No fuel is used to get things started. "Just

light a match," says a DOS employee at the incinerator.
Whoosh! and away it goes. A look inside this Hansel and
Gretel device and one would assume the entire city of New
York could be reduced to a pile of ashes overnight. The heat
is intense. These 200-foot stacks create such an upward
draft that, upon peeking in, one feels as if arms and legs and
ears will be sucked off the body and go flying into that fiery
furnace, thence on to oblivion.

Even after subjecting garbage to 1600° heat and licking
flames, which reduces the garbage volume by about 80 percent, the residue is still recognizable. Aluminum trays,
cans, newspapers scraps. A Lark cigarette pack practically
intact floats by on the conveyor belt after a dip in the water
tanks that cools the hot garbage before it is dumped in the
barges to be carried off to the landfill.

The Brooklyn Incinerator burns six days a week. And on
the seventh it rests while DOS crews come in and shovel
up 35 tons of fly ash by hand.

The largest landfill in the city—and, obviously, the
world—is the Fresh Kills site out in Staten Island. It is 3,000
acres of garbage along with rats, mice, pussycats, fifteen
kinds of sea gulls, and a ninety-three-year-old bird watcher.

Fresh Kills is a strange place, an ethereal sort of nether-
nether land. Part of it is like the ash heaps of *The Great
Gatsby*. Vast, forlorn, endless, and seemingly devoid of life.
The rejected residue of civilization stretching endlessly to
the horizon. To visit Fresh Kills is to have a vision of death,
for it is indeed the final resting place for the unwanted
remains of the biggest city in the United States. One cannot help but view Fresh Kills with incredible, ineffable
sadness. For if this is all there is to life—the world's biggest
garbage dump—what are we all doing here? What difference does it make to work and earn and live and love and
experience joy and pain and sadness if this is where it all
ends up: in a garbage dump with fifteen different kinds of

sea gulls and a smell reminiscent of milk gone sour. Of life somehow gone sour.

A discarded funeral wreath, a doll with arms outstretched to some unseeing child, a matched set of lawn furniture, a chaise longue, and a chair with green and white webbing on an aluminum frame. (These will later be salvaged and, along with an empty wooden cable spool, serve as an oasis for the Fresh Kills staff.) A black and white plastic raincoat. A man's black sock. Next to it, a lady's nylon stocking.

The garbage—25 percent of it burned in incinerators, 75 percent of it raw and rich and ripe with life—is barged to Fresh Kills, some 11,000 tons a day. From the barges it is loaded onto wagon trains, usually two wagons hooked together, that truck it on out to its selected dumping site. Once loaded, the wagons waddle over the landscape and disappear into those 3,000 acres as effectively as if they had been erased by some Jolly Green Giant lurking just around a bend in the garbage.

"Cars really take a beating out here," says the DOS man. "All that gull shit." Indeed, they are a low-flying squadron. Thousands of them, swooping through the air, turning and banking sharply, then dipping down and settling in for a good feast. "Why not?" the DOS man asks. "This is quality garbage." He makes an observation: "There's a difference in garbage, you know. Some Japs were here once and they noticed it right away. We have a different quality of garbage. Higher quality."

Periodically fires break out in that 3,000-acre stretch of landfill. The DOS hustles out an earthmover to go out and bulldoze the flames into submission, burying them under a pile of garbage and smothering their enthusiasm.

The rest of Fresh Kills, in those sprawling 3,000 acres out beyond the unloading docks, is strangely beautiful in its own way. Close in, there is an overpowering feeling of garbage. It is both rife and ripe. Garbage that is easily recognizable. Farther out, where the garbage is older and more firmly integrated into the landscape, there is grass. There are bushes, shrubs, trees. Summertime in Fresh Kills is a time of flowers and birdsongs. A flourishing spontaneous vegetable garden runs rampant around the landfill. Some of the best tomatoes and squash and pumpkins in the area are grown and picked at Fresh Kills. Out go the vegetable scraps—seeds and sprouts—and those lucky few that end up in the world's largest natural compost heap flourish. Come fall, offices all around City Hall are festively decorated with gourds and pumpkins harvested at Fresh Kills. Pheasants and quail scurry through Fresh Kills' underbrush in the fall, creating still another problem for the DOS: hunters who attempt to invade this lush, protected game preserve.

"Fresh Kills turns me on," Jerome Kretchmer enthused one day. "If it happened in New York it's out at Fresh Kills. You can stand on the shore on Monday morning and watch the barges going out to Fresh Kills with the garbage. And you know what went on in New York over the weekend. There are dead rats and dead cats. Wow, man. Whatever's going on in the city is going out to Fresh Kills. Packages, boxes, cartons from fancy stores, dress scraps, Kotex boxes. You can see it all. What we wasted. What we used. In our lives." He even took his seven-year-old daughter's class on a field trip to Fresh Kills.

He sighs. His eyes glaze over. Is his breath really becoming shallow? "Everybody should see Fresh Kills before they die." Die happy—see Fresh Kills. "I use every excuse in the book to get out there." He loves to tramp through Fresh Kills, master of all he surveys. Bell-bottoms flopping, wide tie flapping, mustache quivering. Fresh Kills is the Parthenon to Kretchmer. His temple of garbage. It is his Chartres, his Versailles, his Hadrian's Tomb all rolled into one and

smelling to high heaven. Jerome Kretchmer would probably take one look at Saint Mark's in Venice and compare
its pigeon count to the Fresh Kills gull count. Fresh Kills
would win, hands down.

"It sure has changed out here," said one worker who has
been there for years. Fresh Kills, which opened in 1948, is
built up on low-lying swampland that once proliferated in
the Staten Island region. With both increasing garbage production and rising land costs, the combination of both
seemed the best of all possible worlds. The best means to
two ends: get rid of the garbage, get some new land for real
estate development. "Why, there used to be fresh natural
springs over there." He threw his arm out over hundreds
of acres of garbage. Natural crab beds once flourished in
the area. Now they too are gone, buried under hundreds of
thousands of tons of garbage. He seems sad at the thought
of the obliteration of those swamplands and the concurrent
loss of wildlife. But, he figures: "You gotta put this shit—
oh, pardon me—this stuff somewhere. I guess."

But face it: New York City has got to find a way out from
under all that garbage. By the end of 1972 the city was down
to five proposals. One would have merely extended the life
of Fresh Kills. One was a modest 150-ton-per-day composting operation. A third was a doomed plan to rail-haul garbage to upstate landfills, which was being fought on political grounds (nobody wanted New York City's garbage).
Only two remained with any glamour: a plan to burn garbage for energy that would be sold to Con Ed for heating
and electricity; a 1,000-ton-per-day Landgard system that
would simply dispose of the garbage, via pyrolysis, with
little or no thought for resource recovery (unless money
could be gotten from the recently passed Environmental
Bond Issue to add metal and glass recovery processes to the
Landgard system). Somehow, through it all, Kretchmer
remained hopeful. "It'll happen, it'll happen," he kept re-

peating over and over, like a litany. One could only hope
so. The prospect of 29,000+ tons of garbage piling up each
and every day is not encouraging.

New York City's garbage is also the country's most political garbage. Garbage has become, in the past few years,
one of New York City's hottest political issues. For a city
to let her garbage all hang out is a little like taking the
family skeleton out of the closet and propping it up on the
front stoop. When policemen are on the take, who's to
notice? When housing authority grafters are at work, who's
to notice? But when garbage doesn't get picked up—people
notice. (Especially if they are downwind.)

Mayor John Lindsay learned that lesson when he began
planning to run for his second term in 1969. It was then he
realized the political potential of garbage. "Garbage has
become the 1970s police issue," said Jerome Kretchmer.
Indeed it had. Lindsay needed the endoresement and support of John DeLury's 10,000-plus uniformed sanitation
men, and to get it the usual political bargains were struck.
Lindsay promised to hire 1,000 new sanitation men, and for
that he not only got the support of the Uniformed Sanitationmen's Association, he got cleaner streets: all over the
city collections increased and sanitation men hit the streets
with brooms and shovels on an overtime basis. All just to
clean up New York City in time for election. (This, of
course, was after that incredible nine-day garbage strike in
1968 when nearly a quarter of a million tons of garbage
mounted up uncollected on the sidewalks of New York.)
Fortunately for both the mayor and the city, DeLury came
through. The streets were cleaned up, the mayor reelected.

As usual, it was union chief John DeLury who made the
most sense during that political fight for possession of New
York City's garbage. "People are slobs," he said, quite simply. "New Yorkers just don't care. Cleanliness is not a way
of life here. I don't give a damn what the sociologists

say—the problem is that people don't care, and that goes for the guy who throws stuff out his car or truck window and the storekeeper who leaves crap in front of his store. Sure, there are nice guys in every store and proud of their place. But they don't last. They give up."

Part of it is union troubles. DeLury has so constituted the union and its work rules as to make it almost impossible to shift men around because of seniority. Since the majority of sanitation men are white, the task of sending them into the already hostile ghettos, where garbage has gathered over the decades, is troublesome, to say the least. Ghetto residents see garbage and garbage men as one and the same —especially if the garbage man is white. Part of it is sloppiness and slipshoddiness on the part of individual sanitation men. Part is, of course, due to the incredible increase in garbage production in New York City. And part of it stems, naturally, from the whole range of sociological problems present in, in particular, the ghettos. Absentee landlords, lack of effective superintendents in the buildings, lack of garbage cans, and fear of residents even to go out in their hallways, much less out on the sidewalks with a load of garbage. Abandoned cars make it impossible for mechanical sweepers to clean the streets. Vacant lots abound with garbage collected behind fences where it stinks and molders for years.

Lindsay himself, off on one of his celebrated walking tours, found himself confronted continually with those vacant lots full of garbage. Indiscriminate dumping on the part of supers, vandals, and residents. Professional littering by private garbage haulers who sneak up in the dead of night and dump in a vacant lot rather than take their load down to the docks for barging to a landfill. "I get madder than hell," he stormed. He was all ready to go on television and denounce the dumpers for their sloppiness and disregard of the rights of others to live in a clean environment.

He was advised not to, with aides reasoning that such an attack would provoke widespread resentment of him in slum areas.

Jerome Kretchmer, early on in his days as head of the Environmental Protection Administration, learned the lesson of politics in garbage the hard way. The ebullient administrator had called a news conference but quickly canceled it when he learned the implications of what he was going to announce. Sources speculated that he was going to announce a cutback in collection services in many single-family-dwelling neighborhoods (translation: white) in order to send more men and crews into high-density areas (translation: Black and Puerto Rican). According to sources within the EPA, the news conference was called off because of the "unpleasant political repercussions" from the affected single-family neighborhoods that were so important to Lindsay's reelection.

Out in Brooklyn residents of the Brownsville section— called "Bombsville" by Lindsay—charged that sanitation trucks had missed five of the scheduled six pickups in one week. Then they took matters into their own hands. Literally they picked up their own garbage and spread it around their streets to dramatize their plight. In order to get the trash picked up, however, san men had to be called in from Queens routes, thereby cutting back pickups in that area from three to two times a week.

With the possible exception of a strengthened chain of command at the block level and increased collection in high-density areas, the situation has not changed all that much. Fines are still low, prosecution for large-scale littering is practically unheard of, summonses[3] for inadequate facilities for garbage storage at the household level are still

---

3. You cannot give a summons to a guy who isn't there. To an absentee landlord. To a nonexistent super.

low. The backlog of summonses in the courts still remains. Kretchmer and the EPA are attempting to get an Environmental Court established that would handle all garbage summonses, thus taking them away from the already overburdened courts and putting high priority on them.

Nor is it an easy task being a sanitation man. Louis Campisi, Eddie Vincelli, and Alfred Izzo are three New York City sanitation men. Their beat: the South Bronx. If prizes were awarded for garbage, the South Bronx would win. The South Bronx is the Miss America of garbage, competing twenty-four hours of every day. She is Miss Congeniality, Miss Talent, Miss Inspiration. The works. She has everything: apartments, hospitals, schools, industry, businesses, warehouses, markets, elevated trains, expressways, housing projects, Yankee stadium. Messrs. Campisi, Vincelli, and Izzo are there to handle her. In so doing they walk about fifteen miles a day, in all kinds of weather, and lift nearly four tons of garbage. Off the sidewalk, over to the truck, up over their heads and into the hopper, back over to the sidewalk to put the can down. It is more than most of us do in a day. (Maybe even in a lifetime.)

The day I spent with these three men was miserable. The rain was coming down in sheets. I was not quite as well prepared as the Three Musketeers I was following around —they were each encased in slicker suits from stem to stern, plus much sturdier and more waterproofed shoes than my Army–Navy combat boots. There I was, with my London Fog raincoat, my umbrella, my very wet blue jeans, and those combat boots, which got heavier and heavier as each minute slogged by. Plus my steno pad, whose green lines started to run immediately.

The rain might keep some of the New York natives indoors—junkies and rats and dogs—but it had not kept the groundswell of garbage down. In some places the garbage cans were ready and waiting, like sentinels outside the wall. In other places the garbage spilled into the streets, obscene jewels in a tarnished setting. The rain didn't help matters. A woman coming out of a tenement sets the bag by the can. By the time the sanitation men get to it, the bottom is out of it and its contents—egg shells and coffee grounds and plastic bottles and papers—are spread all over the sidewalk. Across the street, the airmail garbage deposit is being put to use: three bags of garbage arc through the air from a fifth-floor window, landing with a splash! and a splat! in a refuse-ridden vacant lot down below, joining a long list of previous deposits.

Three teen-age girls accost me on a corner, demanding to know what I'm doing. "What're you fucking around here for?" they query. My answer about garbage does not satisfy them. "Garbage—shit," says one, jostling me up against—appropriately enough—a garbage can. "Who gives a shit about garbage." Just to prove her point she kicks the can over, and its contents spew into the gutter. "How come?" I ask, gesturing at the garbage. "Kiss my ass," is the reply. The bad vibrations are like electrical charges out of *Frankenstein*. Salvation, as usual, comes in strange forms: this one a trio of cool dudes bopping into a corner candy store. My charms are, fortunately, both secondary and transitory, and I am left standing in the slicing rain next to a gutter full of garbage.

The rain by this time has made a hobby of me. I am soaked right down to my skin. I think about cops riding around in patrol cars. I think of mailmen who at least can get into buildings and have the comfort of realizing that other people know they go out in snow and rain and gloom of night. Sanitation men are out in it—with no break and no slogan chiseled in stone—year in and year out. Summer heat with its concomitant stinking garbage. Winter rain and snow and wind. No season is a good season for garbage. Summer? "The natives get restless," said one of the men.

"Their houses are hot, so they sit outside. They drink a lotta beer and throw the cans around. Then they get a little stoned and throw their weight around. The kids don't have nothin' to do, so they get inta trouble. Summer? Awful. Stinks like hell, too." Winter? No better. The garbage might keep a little longer, but try walking those fifteen miles and toting those four tons in freezing temperatures complete with snow, ice, and salt.

And the hazards of being a garbage man?

"Dope addicts," suggests Campisi. "You never know what mood they'll be in." Hostile, nodding, high. Junkies have been known to attack sanitation men just because they happened to come by and the junkie was getting desperate.

"Just last week I had my tires stolen," says Izzo. "Outta my trunk. Brand new. I didn't even have a chance t'use them. Can you believe it." Assistant Foreman Peter Barnett parked his car for five minutes the week before—and his battery was clipped out. Burglars once broke into the district's storefront headquarters and stole all the sanitation men's uniforms.

How about dogs? The South Bronx has packs of wild dogs that roam the streets. They are street-wise dogs indeed, operating very much like street gangs. There are sentries, lookouts, hit men. They are constantly attacking police, housing project patrolmen, and—you guessed it—garbage men. "Me and another assistant foreman were checking out a vacant lot," recalls Foreman Barnett. "Out came a German shepherd and his gang after us. We just got outta there in time."

"What about rats?" Vincelli asks. Just as ubiquitous as the absentee landlord to the slums of the South Bronx are the rats. They are big and bold. "I picked up a garbage can the other day. Out jumped a rat. I put it down. Nothin'. I kicked it. Out jumped another one. I kicked it again—out jumped Mr. Number Three." The rats are all over, scurry-

ing through the garbage. Up a sanit man's sleeve as he lifts
the can off the sidewalk, past his face as he dumps it into
the hopper, down his leg as it makes its escape. One remembers with anguish the time when Senator Edward Kennedy
and others were trying to get the Rat Abatement Bill
through Congress, a paltry $4 million allocated to try and
fight the rats in all our cities, not just New York City
(where, it is estimated, there is at least one rat per capita).
It was a time of hilarity on the august floors of the American Congress. Rat jokes abounded. Not surprisingly, the
bill failed. So garbage men—and tenants—in the South
Bronx and elsewhere still have to put up with rats in their
garbage, and in their homes, and in their children's cribs.

We have taken lunchtime refuge from the rain in the
local sanitation headquarters. The rain is pelting down
outside. The coffee, brought in from the bodega next door,
is as steaming as we are as we dry out. Izzo remembers a
fire. "Me and two fellas, we were working this street, see.
We see smoke so we run into this building and take these
kids out. We get 'em next door and call an ambulance for
them." He shakes his head. "Nobody even thanked us." He
pauses. "But the kids were OK."

All agree that night work is the most hazardous. "I had
to fight my way outta one block," Izzo recalls. A crowd had
gathered; it got surly, then it got nasty. Bricks and bottles
and rocks started crashing down. Who knows what trig-
gered it. "I just wanted out." Izzo just wanted out. Daytime
is no picnic either, though. One sanitation crew got caught
in the cross fire—as did area firemen—between the cops
and some snipers. Fortunately there was a demolition team
at work and everybody took shelter on the sidewalk under
some heavy scaffolding. "You don't think about these
things when you think of garbage men," says Izzo.

A pair of gloves lasts a month, if you're lucky. "Old cans
cut your hands. People don't care what they put out there,"

says Campisi. Doctors don't dispose of hypodermic needles properly. The garbage man picks up a plastic garbage bag filled with trash, gets punctured by a needle. "Who knows what's on it?" Handling four tons of garbage every day, a guy doesn't spend time going through it to sort out the problems. "I was workin' with a guy once, he threw a load in and a spray can exploded," said Izzo. "Got him right in the face. Hadda rush him off to the hospital."

But beyond junkies, beyond riots, beyond all the other urban hazards that litter the path of the New York City sanitation man, there is traffic. All day long, New York City sanitation men are exposed to New York City traffic. Anyone who has ever driven in it—or ridden in it or tried to walk through it—knows that the Normandy invasion was safer than trying to make it through New York City's overcrowded streets. One garbage man, late of an afternoon, was emptying a can into the hopper of his truck, which was headed west. Down the street came a car, headed west into the red-hot setting sun. The car slammed into the back of the garbage truck. The conveyor belt that carries the refuse into the body of the truck was still running, grinding up everything in its reach. The sanitation man, his legs and pelvis by this time smashed by the car, was lucky to hold on long enough before passing out. The conveyor belt got turned off just before he would have pitched in and been gobbled up. He lived, but barely. A sunrise victim wasn't so lucky: he died at thirty-two, leaving a wife and three kids.

"Remember John?" one of the men asks. "He got caught between a brick wall and a delivery truck." He died the next day.

"Don't you get depressed?" I cannot help asking what, to me, is an obvious question. Barnett shrugs. "What's to get depressed? You get up in the morning, you shave, you come to work, you do the best you can." Meanwhile, the garbage is already being created again. Some services see an end to things and a beginning to things. A fire is put out. An arrest made. Sure, there's always another one—but somewhere, someone else. The garbage? Stay tuned. Same time, same station, same cast of characters.

The economic crunch has caught the sanitation man just as it has every other uniformed city employee. In fact, it is nearly *de rigueur* for cops and firemen and sanitation men to moonlight. Campisi—he's only got one kid so he can go home at night and relax in front of the tube. Izzo? "I got a kid in college." His son, at Saint John's University, studying "to be a teacher." Izzo? "I drive a cab." He laughs. "I go out after work to be my own worse enemy. I could kill myself!" Of all traffic, cabs are the biggest hazard to sanitation men. Vincelli? He's got three kids and perhaps the most unique after-hours job: he plays drums and sings in a dance band. "We're at a real swank joint in Flatbush this weekend," he announces. "The 'El Caribe.' " His group's specialties: cha-chas, lindys, fox-trot. "That's what people want."

The city of Los Angeles, with a population of 2.8 million as compared to New York City's 7.7 million residents, collects only 1.2 million tons of garbage a year. L.A. can get together only 4,900 tons a day compared to New York's 29,000+ tons.

Perhaps what is most appalling to New Yorkers, viewing L.A. garbage from a distance, is the lack of interest Los Angeles gives her garbage. It goes unnoticed for days on end: most areas get garbage pickup and delivery only once a week. (This is increased to three or five times in big apartment complexes, but that still doesn't help the L.A. garbage ego when compared to the seven times a week some areas of New York City are claimed to be treated to.) "It just means I can keep it at home longer," says Mrs.

Gordon Hazlitt, wife and mother. There is probably something to be said for that stay-at-home approach to garbage, but New Yorkers can and do take pride in the fact that they spend so much time both producing and attending to their garbage. (New Yorkers have been overheard pointing out that whereas they have some 11,000 garbage collectors, L.A. has only 900 of them.)

Because of the air pollution problem in Los Angeles incineration is no longer a way of life for its garbage, which is relegated instead to abandoned gravel pits and unused mountain canyons. This, of course, irritates the environmentalists, who look to canyons and arroyos and gulches and gullies as something other than the last home for Los Angeles' garbage. But considering both the tacky amount of garbage and the tawdry treatment given it—imagine ignoring it except for once-a-week pickups—one is inclined to consider even an abandoned gravel pit as being somewhat better than nothing.

Interestingly enough, Los Angeles—Motown West—did not introduce garbage trucks until 1915. Prior to that year the city's garbage was picked up and delivered to dumps by mule carts. Even after 1915 the old habits died hard: the last mule left Los Angeles in 1934.

To think of San Francisco is to think of Nob Hill and Russian Hill and Steve McQueen careening through the streets. San Francisco is fresh air and sunshine and flowers and, to some, the most incredible garbage collection in the world today.

Unlike New York City's municipal collections, done—or undone, as some critics charge—by city sanitation men, San Francisco's garbage is collected and disposed of by private companies. Two of them, in fact. Sunset Scavengers Co. and Golden Gate Disposal. There are those who point to this private treatment of garbage as the best way of handling the situation. This is a highly romantic vision, infused with the whole mysterious mystique of Bagdad-by-the-Bay. Legend has it that Sunset Scavengers—responsible for picking up two-thirds of San Francisco's garbage—roll through town in their red and white trucks working like demons. Running from one garbage can to another, pausing for barely a breath and never for lunch or coffee, singing snatches from *Bohème* and *Traviata* as they go. An eight-hour day turns into a six-hour day, all because of private enterprise. And for this service, householders pay only $2.50 a month for the privilege of having their garbage collected by a man with a song not only in his heart but on his lips. It all makes for happier households and happier garbage.

San Francisco's garbage was originially collected by Italian immigrants with horse-and-wagon equipment who operated as junkmen, picking up refuse, sorting it, and selling it to secondhand dealers. Soon there were upward of 116 companies competing for San Francisco's garbage. This got a bit too complicated—even private bureacracy is hard to deal with—so in 1932 San Francisco was divvied up into ninty-seven garbage districts with, by then, some 36 companies vying for contracts. Private enterprise being what it is, these contracts were soon in the hands of two companies, Sunset and Golden Gate, who now have a monopoly on San Francisco's garbage. Both companies operate on essentially the same basis, with many company employees holding stock in the company. Worker-stockholders get wages along with their shares in the company. Some San Francisco garbage men earn upward of $16,000 under this setup. In addition to owning part of the company, each driver has a vehicle he is responsible for—maintainance, upkeep, cleanliness—plus a fixed territory where he enforces payment, haggles about special collections (stoves, refrigerators, etc.), and lowers the broom on delinquent accounts.

So there they are, San Francisco's legendary garbage men. Running from one can to the next, up steps after cans, ahead of trucks to get things moving. If a garbage man in New York City ever ran he would (1) probably be shot by an off-duty cop who mistook him for a purse snatcher or (2) be shot by his fellow union sanitation men for setting a bad example. Productivity sometimes seems to be illegal in New York City. What the S.F. san men have done, however, is make their own system and then beat it. They are paid for nine-hour days, expected to work only eight (with one hour for lunch), but make do in six.

Up Nob Hill, down Russian Hill, in goes the garbage, out comes an Italian aria. Not that Sunset Scavenger Co. and Golden Gate Disposal are free from verbal garbage. Some customers figure that monopolies make for shoddy service and they sing out to the tune of thirty complaints per week. Leonard Stefanelli, gregarious president of Sunset Scavenger Co., figures he gives the residents of San Francisco a good garbage deal. "Everybody assumes we'll be there," he has said. "And we are. It's like your electricity working. Only we've never had a power failure. We've never even blown a fuse."

It was charged in one magazine article that San Francisco gets her 2,000 tons of garbage a day picked up for about $20 million. That same article charged that New York's 8,000 tons goes for a high of $390 million. This tends to make New Yorkers view the San Francisco scene bitterly. Oh, distortions and distractions: New York's Sanitation Department handles 29,000 tons—not a paltry 8,000 tons—a day.

The garbage really hit the fan in New York in mid-1971 when a report by Deputy Mayor Timothy Costello began circulating in and around City Hall. Costello's report urged that some private pickups be instituted, claiming that it could cut the cost of garbage service in New York by one half. This, of course, hinted at gross inefficiencies and mismanagement and corruption and laziness and everything else in the Sanitation Department. Which, to some extent, is probably true. But when Costello's report began trotting out statistics on how private carting in New York is cheaper by half, the Department of Sanitation fought back. First they pointed out that private carters in New York, who pick up from businesses and larger private institutions, have a much easier time of it. For one thing, they make fewer stops since the volume they handle at each stop is bigger. For another thing, they are not responsible for street cleaning and snow removal and hence are not plagued by abandoned cars and litter and what have you. Beyond that, they do not have to contend with absentee landlords, vanishing supers, and all the other sociological conditions that plague both residents and garbage men in New York City.

Even *Solid Waste Management*, a trade magazine hardly known for its kindly view toward New York City's Sanitation Department and the sanitation union, conceded that New York City had a particular garbage problem. "In East New York, on a recent day," said an *SWM* editorial, "a sanitation task force augmenting regular crews swept through the decaying blocks of the area and, despite the concentrated effort, refuse flowed back onto the sidewalks, gutters and in the streets the following day with depressing regularity." Private contractors who handle the massive gobs of garbage (much of it clean and nonputrescent) that flow out of, say, the 46-story Time-Life Building, or the 102-story Empire State Building, or even the entire Garment Center in midtown Manhattan, are not caught in the crunch like the city collector servicing the endless decaying 5-story slums of Brownsville and the South Bronx and Bedford-Stuyvesant and the Lower East Side.

It is an interesting battle, private vs. public collection.

Meanwhile, like ol' Man River, the torrent of garbage just flows along.

Not that America is alone in her fight to improve the quality of garbage. Not at all. In times of need our friends —our true friends, that is—do not desert us. They stand by us, imitating and emulating, showing the world they are truly with us, sharing our glories. In this case, garbage. Take Japan, for example. In the past year, some 233 extra dump trucks have been pressed into service to help Tokyo's fleet of 3,069 garbage trucks cope with the products of their newly found affluence. Tokyo's normal daily output of 12,500 tons of garbage now includes television sets, refrigerators, radios, furniture, washing machines, and those familiar stacks of newspapers and magazines. Tokyo is truly coming of age in the garbage generation, spurred on by an industrial boom and heavy advertising to encourage consumerism, plus the usual population boom. Growing affluence has shrunk the secondhand market. Men who used to pay for the privilege of removing old newspapers and magazines, or who used to trade them with householders for toilet paper, now have to be paid to cart surplus papers away. "Some of the television sets need only small repairs," said Norio Koike, a section chief in Tokyo's Sanitation Department, one day as he mused at the quality garbage Tokyo was now producing. "But many people people are getting color sets and they don't want the black and white ones anymore." Hence—the quality garbage heap.

And Japan, like America, is also running out of display areas. A third of the garbage is burned, and local opposition prevents the construction of more incinerators. The remaining two-thirds is put in landfill in Tokyo Bay, but in 1967 the major dump site filled up. Tokyo named it Dream Island and moved on to smaller sites, which will be all filled sometime this year. Jerome Kretchmer, on a whirlwind tour of Japanese garbage disposal sites, was taken around the Dream Island complex. He was impressed, undoubtedly thinking thoughts of Home Sweet Home and Fresh Kills. The problems of garbage. Tokyo, too, is facing a garbage crisis of monumental proportions.

Despite the Herculean efforts of, say, Tokyo, it is still to New York City that the world looks for garbage leadership. It is a burden not borne lightly by the New Yorker. Indeed, he bears his burden both seriously and heavily: since 1960 New York City's population rose 1.5 percent And her garbage? Up 42 percent during the same time.

IX

Garbage Men

It goes without saying, where there's garbage there's a garbage man. There is in New York City, this very day, the best-known garbage man in this country. Or as Jerome Kretchmer, administrator of New York City's Environmental Protection Administration, puts it, "I'm the best-known urban environmentalist around." He says this while throwing himself, six feet four, onto the leather couch in his office on the twenty-third floor of the Municipal Building, a gnarled structure in lower Manhattan at the foot of the Brooklyn Bridge. Everybody knows Jerome Kretchmer, or so it seems. One has visions of kids bartering Kretchmer cards into their stack of baseball trading cards. Sending in cereal box tops for an autographed picture of Jerome Kretchmer. Lines of squealing girls who, when he appears, burst into tears and faint dead away into their bobby sox.

It's not quite that bad yet. The city has not yet reached Kretchmer Crisis proportions, although a typical day in the life of the administrator might make it seem so. For his is a round of routine (well, for him, anyway) office work, flying dashes in and out to make speeches before some civic club or merchants' association or school group or any motley collection of outraged citizens, guilt-ridden industrialists, or visiting environmental scholars. He takes on the supermarket interests, the packaging industry, the manufacturing men. They hate him. He loves it. "I really bugged the shit outta them," he said after flying up to Grossinger's in the Catskills to lay his line on a meeting of supermarket managers. "You shoudda seen them. They were ready to throw me out." He loves it.

His eminent domain is the Environmental Protection Administration, one of the ten superagencies created by Mayor John V. Lindsay for New York City. Within the EPA are air and water and noise pollution divisions. Also contained under the EPA umbrella is the Department of Sanitation, that doughty bureaucracy responsible for picking up that monumental donation New York City makes each day to the nation's solid waste stream. New York City's contribution alone makes it less a stream than a flood. Kretchmer is just one finger in the dike, but it's an important one. In fact, for eight months after the departure of Griswold Moeller, New York City's fourth sanitation head in as many years, Jerome Kretchmer was also acting sanitation commissioner. Garbage is the real nitty gritty to Jerome Kretchmer. Something he can deal with. "Garbage is *real*," he says. It has life and body. Oh, boy—does it. New York is not only the richest lode in the whole mothering mound of garbage, it is probably the ripest. There is more than just safety in numbers, fellow citizens.

Now Kretchmer's view of the environment is no heady vision of sunshine dappling through stands of primeval redwoods. Of fresh waters gurgling through natural gorges. Of dew-laden flowers at sunrise and trilling bird-songs at sunset. "Who gives a shit about mountains?" Kretchmer asks. "I don't, that's for sure. I mean I'm a city guy, right? I take my vacations in cities. I love cities. I'm not into rivers. I don't get turned on by mountains. I get turned on when I go into the South Bronx and it's cleaner than usual and people are living better. Cities are where it's at, right? Nothing's happening on the mountains." For Kretchmer, his mountains consist of solid waste. Piles of garbage mounting up at Fresh Kills and the Brooklyn Incinerator and twelve other locations around the city.

It is no surprise, then, that Kretchmer came into the environmental jungle not so much as a missionary as a white hunter, out for what he could get. It was a job, he was tapped for it, he was shrewd enough to see its incredible potential—given, of course, his own personal pizzazz and personality—and he was smart enough to take it. And take advantage of it.

"New York is polluted because people pollute it. Detroit and Chicago and Gary are polluted because the economy pollutes them. But New York doesn't have heavy industry. We have cars, electric generating plants, and garbage. But no industrial pollution. New York is a perfect example of the society just having overrun itself, mucking itself up. Somebody's got to reeducate the people of this city, civilize New Yorkers. And to do that you've got to be in the system." Kretchmer is the kind of political animal dear to all radical-chic liberals: he is a politician who *really* cares.

Kretchmer relishes all the publicity he gains as the world's best-known environmentalist. He probably even believes it. It is almost impossible to write anything dishonest about Kretchmer, for if Kretchmer is one thing he is honest and outspoken. "I made *Playboy* this month, didja see? Lookit this." It's Kretchmer over in Brooklyn, with Manhattan barely visible through some dank, low-hanging man-made clouds. His foot is firmly planted on a wooden chair. "I found this chair floating in the river and fished it out. Isn't it terrific?"

Obviously, then, Kretchmer's vision is a little more basic than most environmentalists'. When Kretchmer sees a stack of newspapers at the recycling center, his mind's eye does not conjure up visions of seventeen leafy green trees per ton of newsprint. Kretchmer simply sees a ton of newsprint that the Department of Sanitation doesn't have to haul away. He doesn't see that ton of recycled paper as nesting places for the birds of the air. He sees it as a ton of paper that didn't make it to one of his already overburdened incinerators where it would produce fly ash and other particulate matter. He sees it as yet another ton of garbage that didn't make it into his already overcrowding landfills.

Environment? "Shit—it's right here," he says, gesturing around him wherever he happens to be. It's the street, the subway, the air. It's noisy jackhammers and dirty gutters and decaying tenements.

Kretchmer aide John Leo walks in waving a recent copy of the *Village Voice*. "They gave you credit in here," he says, referring to an article on talcum powder and cancer. "The bit about blowing the whistle on asbestos." The article linked talcum powder—which is primarily asbestos—to a high incidence of vaginal cancer, citing use of bath talc and talc on diaphragms before insertion. "Well, here we are," Kretchmer noted, "protecting the vagina interests again."

Jerome Kretchmer got his job through politics. As state assemblyman from the Upper West Side in New York City, Kretchmer had proved himself both an able politician and a deft manipulator. "I made the right enemies," he once said. There he was, the cowboy from the West Side, shooting from the hip into the conservative barroom that is the Albany legislature. He rode into town on the 3:10 from Manhattan, dressed in boots and wide ties and a South-of-the-Bronx mustache that flowed like red wine at a New York party. There he was, taking deadly aim at such subjects as divorce reform, educational reform, abortion reform, drug addict treatment and rehabilitation, low-income housing. He took a wide liberal stance and kept shooting with all barrels.

By 1964 he was already making the right decisions. He was one of the earliest supporters of Robert F. Kennedy. He worked on Kennedy's senatorial campaign in that year, and when 1968 came, Kretchmer ended up out in Gary, Indiana, for two months, doing grass-roots political organizing. "Gary, Indiana! I loved it! What a city!" Kretchmer fairly *kvells* at the thought of Gary, Indiana. "It's so ethnic! It's alive. I loved those people. Those steel men. They're terrific. They liked me, too. They really related to me. I'm a regular guy to guys like that." Indeed he is. He not only sounds as if he came right off the street—his nasal tones filtered through his ethnic nose—he looks it. Kretchmer's face is like a face designed by Hamlin Garland—full of main-traveled roads.

Kretchmer was born September 15, 1934, in the Bronx, the son of a Jewish laundry-truck driver. He attended the Bronx High School of Science, one of the best schools in New York City. (Grace Lichtenstein, a *New York Times* reporter, quoted one childhood pal of Kretchmer's as allowing to Kretchmer's ability, but being somewhat surprised at his swift rise to the top, particularly in city and state politics. "He was such a *nebbish*. A guy who wanted desperately to belong to the 'in' crowd.")

Well, Kretchmer grew up and overcame all that. He was a scholarship student at N.Y.U. and Columbia Law School, waiting tables during the summers in the Catskills (where, by the way, he met his wife, Dorothy. He was a waiter, she was a camp counselor. *Marjorie Morningstar* all over again.)

Now the 1950s may have been our last days of innocence, with the most predominant feature being the decisive calm of Ike. The world was smoldering away beneath our feet like a cauldron of hot oil. Topside, however, there was not an oil spill in sight. We closed the circle by wearing a circle pin.

In a certain all-American way Jerome Kretchmer is indeed a child of the 1950s. If he is today something of a joiner —joining in the fight against pollution, noise, substandard housing, inequitable schools—he was just as much a busy joiner back in the Class of '55. His yearbook picture has a copyblock that is so typical of the 1950s it seems to stand up and shout sis-boom-bah. There he was, editor-in-chief of the *Violet*, on the Junior and Senior Prom committee, the elections committee, finance committee, president of his sophomore class, freshman class secretary, chairman of the social affairs committee, co-hazing master, John Marshall Pre-Law Society, Frosh Follies, plus intramural basketball and softball and a string of other *de rigueur* 1950 activities. Kretchmer in his white bucks and salt-and-pepper tweed sport coat and white shirt open at the neck. Kretchmer at a school dance, short hair in a sculptured pompadour. Kretchmer at a school carnival with pie on his face.

Once out of law school and into practice—Olshon, Grundman, Frome and Kretchmer—he was on his way. In 1960 he founded the Ansonia Independent Democrats, then went out and hustled the West Side voters. He got himself elected district leader in 1961 and state assemblyman in 1962. He was an assemblyman for seven years, getting more and more frustrated at the conflict between the crusty old conservative politics of the upstate legislators and his more gutsy brand of politics.

Enter John Lindsay with a vacancy at EPA. The then Deputy Mayor Richard Aurelio mentioned Kretchmer's name; Lindsay figured why not? as did Kretchmer. "Then the mayor called me." Kretchmer, who was by this time thinking of running for the Senate, called his brother in Chicago, "who said I was a schmuck if I ran for Congress. I'd get lost in Congress, he told me." They mulled over the pros and cons and figured it was as good a place as any to build not only bridges, but bases. "So I went back and asked the mayor how much he'd bother me and he said not much, so I said yes." In May of 1969 Jerome Kretchmer became New York City's second Environmental Protection Administrator.

Right now Jerome Kretchmer wants to grow up to be mayor of New York. "Why do you think I took this job?" he asks, as if it should be a rhetorical question. "It's a good place to show commitment. It's a good base. It's safe. Everybody wants to clean up the environment." He just might make it all the way to City Hall, having hitched a ride on the country's biggest garbage truck, for Jerome Kretchmer is the kind of guy New Yorkers can relate to. If John Lindsay came right out of the Silk Stocking district via private schools and townhouses, giving New Yorkers their own brand of sex appeal, Jerome Kretchmer came right off the streets of New York. If John Lindsay exudes a sort of cosmopolite charisma, Jerome Kretchmer is still brushing the Bronx from the seat of his pants. John Lindsay is Hollywood handsome in movie-star sort of way—an aging Tab Hunter. Jerome Kretchmer is the guy next door. His eyes are hooded, his teeth have gaps. When he smiles it's like a basset hound doing an imitation of Terry-Thomas.

Jerome Kretchmer may walk around today, thirty-eight years after his Depression birth in the Bronx, wearing suits from Saks Fifth Avenue and fifty-dollar Gucci loafers from

Italy. But he talks Bronx and lives on the West Side of Manhattan. "I send my kids to public school." And that just might have made a lot of sense to the voters of New York City come the 1973 mayoral election. Kretchmer has enough political savvy to know his constituency: the people of New York. "Cops love me. I'm the only radical they like. That says a lot." In New York City, where the cops figure Lindsay uses them for doormats, that means a lot, too. "Sanitation men, too." A *Time* magazine correspondent corroborates Kretchmer's image: "I was having lunch with him and you know what he did when he finished eating? He picked up the edge of the tablecloth and wiped his mouth with it."

Once every week Jerome Kretchmer gets up at 5 A.M so he can set out at 6 A.M. to make a 7 A.M. roll call at one of the city's fifty-eight sanitation garages. He loves it. They love it. He makes small talk between dodging trucks and brooms and men, then heads for the locker room, where the chitchat roams over working conditions, personal gripes. "They know I care," he says. "It's good for morale." It's also good politics.

Once the inspection is over, the trucks start to roll out. Out they come, dozens of white sanitation trucks with one driver and two guys riding shotgun on the sides. And what awaits them? Human debris, the leavings of our lives. If anyone is depressed in the environmental crisis now facing this country it should be the garbage men. For it is true that no sooner do they get to the end of the street than it has already started piling up back at the other end. Being a garbage man must be the most frustrating occupation in the world: there is no end to it.

When it comes to garbage Kretchmer is not only a bureaucrat, but something of a sociologist. As he drives back through the South Bronx, one of the most depressed and depressing areas of New York City, the debris is already spilling over from trash cans to the sidewalk to the street. Vacant lots are steaming and rancid with trash. The buildings look as if they are World War II rejects. "People ask why we don't pick up the garbage here," he says. "We pick it up seven days a week. There's more to garbage than garbage men, you know. Look at this, just look at this," Kretchmer says, gesturing around him. "What a fucking mess. It's not the fault of the people who live here. Look— the housing is decayed. They're trapped. The real problems are these: the buildings are decayed, there are no supers, the people have no control over basic services. Everything kind of spins off of that and magnifies. It's not the people's fault it's filthy up here. It's not the garbage man's fault. When you're worried about how you're gonna live from one day to the next—with junkies and no heat and no hot water and peeling paint your kids eat off the walls and get lead poisoning from—it's hard to get upset and talk about dirty streets. Part of all this shit is just plain fear: people use airmail because they're afraid to go down and put their garbage in the cans. It's all wrapped up in an uptight little ball. It's a vicious circle. Incestuous."

New York's problems are monumental, there is no denying it. With a population of 7,771,730 and a land area of 320 square miles, New York's population density averages 25,000 people per square mile. In Manhattan that figure spirals dizzily to 66,000 people per square mile. Compared to Manhattan the borough of Richmond is a veritable wasteland, with only 4,000 per square mile. Tokyo, on the other hand, has its 10 million people spread out over 800 square miles.

And Kretchmer is right: when there are more basic things to worry about, garbage and clean streets come low down on the priority list. "Take furniture," Kretchmer said a few days later, sprawled out on the couch in his office. "This is a good example of how society fucks up. Poor people buy cheap furniture on time and it breaks, so they throw it away. Whole bedrooms go out on the street. The damned things don't last. There are no guarantees. Nobody standing behind this shit because they're all out to make a buck—the manufacturers, the store owners. And these people can't sustain the energy to get things repaired. My wife can. She can get on the phone—like every other middle-class housewife—and yell until things happen for her. A poor woman, worried about a thousand other things, doesn't have the energy level to do this. And keep at it, because that's what it would take. Lots of bluff and bullshit. They're too busy surviving."

Thus Kretchmer is also enough of a liberal to know that, just like the outdoors, everything in the city is likewise a circle. And just to pick up the garbage isn't enough. You have to start where it starts. And often it starts with poor housing, depressed spirits, dope, second-rate educations, and broken homes. It's a socioeconomic problem, with no easy solutions.

"Take our war psychology: a whole segment of the country believes that we have to maintain a certain level of defense preparedness because if we don't the economy will fall apart. That's bullshit. I don't believe that for one minute. Like, if we took all the ingenuity that we put into developing plastic packaging and put it to use making industrialized housing—quick, decent housing—we'd be a lot better off. We'd be able to house millions of the ill-housed. Instead, we manufacture plastic packaging. . . .

"I think the packaging industry, the drug companies, the television manufacturers, the styrofoam users—these are the bad guys," he says. "The packaging of these industries is symptomatic of their disrespect for the environment. The packages aren't full and the products either don't work at all or don't work for long—the radio that gets tossed out in six months because it's broken or because there's a new one out with seven steam vents on the side. Planned obsolescence is both the villain and the machinery that runs the gross national product."

Kretchmer had his chance with his packaging enemies in 1972 when he addressed a packaging conference and told them right out that he has received no cooperation from bottlers, canners, and packers who have added an "extraordinary mess" to America's garbage problem. "All I want you to do," he told the gathering, "is assume some share of the burden. Until we get some cooperation from people like you, people like me will continue to be angry." Naturally this did not go down well with the assembled packagers. "I don't believe you!" cried Warren E. Craumer, a representative of E. I. duPont de Nemours & Co., jumping angrily to his feet. "Industry has as much concern about the environment as you!" he yelled at Kretchmer, both his voice and his fist shaking.

Kretchmer loved it.

When the New York City Chamber of Commerce sponsored a seminar on recycled paper—what it is and how it can be used—Kretchmer opened the day's proceedings. The meeting was held in the C of C building, an encrusted landmark in Lower Manhattan full of carved wood and sweeping marble staircases and glowering busts. The Grand Hall, modeled after the Guildhall in London, is a soaring affair replete with carved and gilded rosettes on the ceilings, fake skylights, and panels of skies in the ceilings with faint clouds moving across them. "God, this is awful," Kretchmer noted. "Look at all the excesses *here*!" He was introduced to the Chamber of Commerce as a "burr under the saddle of consumerism" and a "gadfly in the service of improvement." The inclination was to scratch.

"I'm here to talk about why it is so important for this environment we are supposed to save here in New York City," he told the group, representing some of the most powerful business interests in the country. "Not to make it [excess packaging] in the first place. Not to get it into the solid waste stream. This whole idea of excess packaging—you pay thirty-nine cents for a package of screws. That's one cent for the screws, nineteen cents for the packaging, and nineteen cents for profits. You don't know what to keep and what to throw away. The Polaroid camera—half the time the picture doesn't come out, so you throw that away, too, all of which is designed to produce business. It's no accident," he told the men. "This stuff is manufactured by somebody."

His solution? "Reuse. Recycle. Break our solid waste back down into its component parts. You know," he said, "our solid waste grows at about the same rate as our gross natonal product. If we have this fantastic growth Nixon wants in order to get reelected next year, we are going to have more garbage. It's that simple: Nixon's running on a garbage platform. Growth equals garbage. This is the most political problem this city and this country faces."

One of the things Kretchmer loves to hate is the car. As far as he's concerned there could be roadblocks thrown up around New York City to keep them out. "But I used to have real affection for cars," he mused one day. "They're so masculine, you know. I used to read *Road & Track* and everything. Now? Now they're an extravagant pain in the ass." Up in the country Kretchmer relies on a '65 Volvo, which "I keep in good repair. It'll do. I'm not trying to prove anything with it. Just get to the store for eggs and bread. Like that."

For the 1971 Auto Show—that annual machismo machine display at the New York Coliseum—Kretchmer and his EPA manned an anti-car booth. "We're attacking the hardest audience," said Kretchmer, who figures the car is "anti-city" and "by far the greatest single source of air pollution by weight in New York City." The booth was a potpourri of snappy graphics and multimedia slide projections and a jazzy twenty-three-page booklet drawn up by Kretchmer's EPA staff entitled, appropriately enough, "The Car Is Anti-City." (It was printed on recycled paper.)

"We're going to have to consider the role of the automobile in the central city and perhaps think about banning them on certain days and at certain times or limiting them to certain lanes and creating bike lanes. I think this is the city's first realization of how evil the car is and how miserable the transportation system is. The goal of transportation at this point is to get you—not to get you anywhere but to get you. To wear you down. To wear you out. A guy gets on the subway in the morning in a pretty good mood. Half hour later he's ready to kill. It's really the perfect example of a corrupt environment.

"I'm for the simple things in life," he sighs sarcastically. "Mass transit instead of cars. Industrial conversion. Why the hell can't Boeing build homes for people? People are very adaptable—but our industrial system is very unadaptable."

But it's hard, being the world's best-known environmentalist. His family has had to fall in line. His wife, Dorothy, uses white toilet paper and biodegradable laundry detergent. And their three kids had to learn the fine art of tin-can stomping. Kretchmer wants to air-condition their new Central Park West apartment (located, interestingly enough, in a building called The White House) but worries about the whole power problem that arises when people do things like air-conditioning their homes, thus necessitating new power plants that pollute the air if they are fossil-fueled or the waters if they are nuclear-powered.

"Face it—I'm like everybody else. I'm greedy. I like to get a couple new suits a year. I like my wife to look nice. I want my kids to look nice. Look—we're all consumers. The thing is, we gotta figure out what's important. Like, we don't use disposables in my house. We don't like them. In the country, we give dinners for forty to fifty people and we use real dishes and real glasses." He is true to his commitment: at a party for Barry Commoner, upon the publication of his book *The Closing Circle,* there were over a hundred people and real wine glasses.

Kretchmer's EPA is a tight-knit staff of very loose environmentalists. It is a coalition of crew-cuts and long-hairs. "I want my people to have fun," he says. Hence Friday nights often bring a round of wine and cheese (using real glasses) to the halls of the bureaucracy. Even those halls at Kretchmer's end of the twenty-third floor of the Municipal Building are different. He painted the walls clean white, replaced the old grim, dim light fixtures with modern round globes. Real plants line the corridors, giving one the feeling of an upbeat Palm Court. Here and there huge white barrels are placed as receptacles for newspapers and cans to be hauled off to recycling centers. Many of the staff offices are painted in warm tones of pumpkin, tangerine, and russet. Dramatic black and white posters of New York City line the halls and doors of the EPA. He shares the floor with the Bureau of Gas & Electricity ("They're the ones with quill pens," said one New Yorker who had occasion to visit the floors) and the contrast is stark, to say the least. "They used to decorate with maps of gas mains here!" Kretchmer cries.

Not that Jerome Kretchmer is universally loved. Mention his name, and words like "politician" and "cheap gimmicks" and "political self-interest" are sure to smatter the air. Kretchmer once went to the Poconos in Pennsylvania to deliver a speech to a management firm's seminar. He took his family along both because, as he said, it was close and because he had been away from home a lot at the time and thought it would be a treat for his wife and kids. His enemies immediately attacked him for being on the take.

Kretchmer responded by sending a check off to the firm to cover his family's expenses, then mused wryly, "I might have a price, but it's sure as hell not a weekend in the Poconos."

The commissioner of sanitation in New York City—the Department of Sanitation being part of the Environmental Protection Administration—is Herbert Elish. The differences between Elish and Kretchmer are striking. Elish is as low-keyed as Kretchmer is flamboyant. Elish is gray suit, Kretchmer a flowered tie. Upon interviewing Elish, then, a person figures he should walk in with a hearing aid, looking for a guy somewhere in the vicinity of the face on the barroom floor. Well—not really. Granted Elish is not as outspoken and brash and irreverent as Kretchmer. But he is not invisible. (In fact, the only negative thing to be said about Herb Elish is his disconcerting resemblance to Richard Nixon. But right there, the resemblance ends. Herb Elish is a human being.)

Herb Elish, like most do-gooders in the world, got into the Department of Sanitation quite by accident. He does not look like a garbage man, nor does he think of himself as one. "I'm an administrator," he says. "I move things around. I allocate men and equipment and services." He was a Washington lawyer, who one day discovered "I didn't like practicing law. It was not a very rewarding way to live my life—counting money. I wanted to do something that had more social value." Fortunately he knew someone who knew someone who knew someone in city government. "I came up to New York to 'have a talk' and met Kretchmer." He became Kretchmer's first deputy administrator in 1970, sanitation chief in 1971. He freely admits he had no environmental background whatsoever and "certainly no background in garbage."

Elish is no boy wonder. His public profile is indeed so

low as to be almost nonexistent. The worst that can be said of Herb Elish is that he is a nice guy. And his heart is absolutely in the right place. Where Kretchmer can worry about industrial conversion and technology and recycling and excess packaging, Elish is tied to the more realistic and day-to-day problems of "picking it up, and putting it down."

That, however, does not relieve him of the task of thinking about where it comes from to begin with. "Kretchmer has this view that somehow business ought to be required not to pollute. Not to produce garbage. Throwaway cans and excess packaging and all that. I don't think that's possible under our present economic system. And I don't believe we are going to change our economic system." To Elish the real problem is "society and affluence." Too much money to spend on consumer goods. "The tax rate in this country is too low. To buy a new car every year: there is something wrong with the basic system that gives people enough money to spend on waste. We ought to spend our money on schools and job training—trying to end the waste of all kinds in this country."

Herb Elish sat in his office at the Department of Sanitation one crisp October day. It is a sweeping wood-paneled office on Worth Street with an air of warm efficiency about it. "This city isn't going to survive," Elish says glumly, leaning back in his chair, his hands clasped over a pencil. "We should take the money from people and redistribute it. God—I sound like Fred Harris. A New Populist! But what I'm talking about is so simple and so basic. It's a bag around our oranges. A new car every year. Our priorities are all warped. There should be more money in the federal budget. You're never going to convince business to change its course as long as people buy what business produces. You can't expect them to be good guys. That's not the good old American way for business."

"You're really dealing with several cities," Elish once said of his chores as sanitation commissioner of New York City. "You have the business areas with great concentrations of people and littering. You have store owners who don't keep sidewalks swept. Then you go to the ghettoes where sanitation problems are more than sanitation problems. It's hard to expect people to come out and sweep their sidewalks and containerize their garbage if there are rats in their building and if their kids eat paint off the walls and get lead poisoning. All you can do is try to make it as good as you can."

Herb Elish is John Lindsay's ninth sanitation chief since Lindsay's swearing-in at City Hall in 1966. Part of the problem is the job itself. There is little glamour in garbage for the average person. Secondly, there is John DeLury, head of the Uniformed Sanitationmen's Association. It is said John DeLury could get rid of God himself if he, DeLury, did not like Him. Fortunately DeLury likes Kretchmer and, so far, Elish. Kretchmer and DeLury are an Odd Couple, Kretchmer's lean and lanky six feet four towering over Delury's short and squat five feet five. But they both speak the same language. In heavy, simple New York tones.

John DeLury, who runs the USA with an iron hand, is built along the lines of Jimmy Cagney: short, barrel-chested, and with a jaw like a tool chest. He looks like a cocky, banty Irishmen who would be more comfortable in bars like The Shamrock or The Blarney Stone. He walks with short, brisk strides and rolls from side to side. DeLury, however, lives and breathes USA. He sits in a swivel chair down at union headquarters, feet barely grazing the floor, behind a massive walnut desk that looks as if he could dive under and be swallowed up in it if he put his finger aside his nose and chose to do so.

It is up to John DeLury to see that the 15,131 men in the Sanitation Department are given not only the fairest shake

by the city of New York but the best. Between John De-
Lury and his chief aide, Jack Bigel, an urbane, chess-play-
ing New Yorker who comes to the USA from his days as
a 1930s intellectual labor organizer, the USA is one of the
strongest and best-organized unions the city has to deal
with.

DeLury is a fighter, a feisty man with "New York"
scrawled all over him. He might wear well-tailored suits
and sit in a big office behind a big desk, but to the garbage
men in New York he's all theirs. "I think John DeLury gets
up in the morning and stands in front of the mirror practicing his New York accent," said one EPA aide. And if
Jerome Kretchmer looks as if he washes his face with Ajax,
John DeLury sounds as if he gargles with it. So seriously
does John DeLury take his job and so serious is his commitment to his men that he spent fifteen days in jail during
New York's 1968 garbage strike for defying a back-to-work
court order. The men didn't go back to work and the city
was paralyzed as nearly a quarter of a million tons of garbage piled up around the city. Fortunately it was winter,
and the stuff froze in big chunks or at least stayed cold
enough not to get any more fetid than it already was.
Mayor Lindsay and Governor Rockefeller threw verbal
garbage at each other as they fought over who'd pick it up.
And put it down. John DeLury was not a very popular man
at that time, except with his men.

While Kretchmer and Elish and the environmentalists
talk about excess packaging and economics and technology,
John DeLury and his men are wading through garbage.
"The people of New York City are slobs," says DeLury.
Well, that's that. "New York has the biggest landfill in the
world: the city streets. How can you talk about recycling
and separate collection and waste avoidance if people
throw their garbage in the streets?" DeLury sees the whole
environmental movement as just another fad, like the anti-

war movement and radical politics and women's lib. "The faddists can always find another fad," he says, "but we sanitation men, the practical ecologists of the city streets, must stay with the almost impossible job of striving to keep New York City clean."

DeLury figures the only way to stop people throwing all those beer cans and candy wrappers and cigarette packs into his streets is to educate them. "Keep at it," he says. "You gotta keep at it. No one-shot deals. No gimmicks. Pound it into them."

While the environmentalists around him are yelling for garbage separation at the household level and recycling, John DeLury just fumes. "Ah, Christ—you can't even get the slobs in this city to *use* their garbage cans. How the hell can you make 'em separate their garbage. And if they don't separate it—bottles and cans and papers and foods—whattaya gonna do—leave it there? Let it sit and stink and get thrown all over? Oh, Christ—be sensible." It is a sensible enough argument.

John DeLury couldn't care less about what causes garbage. He just wants to take it from there: keep it in neat piles and get it picked up. It is an honorable enough goal.

Kretchmer looks at the South Bronx and sees a depressed area awash with garbage and decaying tenements and junkies nodding in the streets. It is a view through the eye of a sociologist who sees people fighting for their very lives. He sees desperate people leading desperate lives, with garbage a very low priority. DeLury looks at the South Bronx with the practiced eye of his mostly white lower-middle-class working-class sanitation men. He sees garbage. He sees slobs. Period. People who don't give a damn. He sees the South Bronx as an area of preferential treatment as far as garbage pickup goes. "We pick up twice a week on Park Avenue," he says. "Seven times a week in the South Bronx. And lookit the difference." Well, maybe, the difference is the difference between the women on Park Avenue who have maids—and the women of the South Bronx who, if they are lucky, *are* the maids on Park Avenue. Maybe it's just the difference between having a super and enough garbage cans, and not. At least, that's how Jerome Kretchmer sees it, at the beginning. DeLury sees it at the end: garbage.

At least they're both seeing it.

X

Gourmet Garbage

There are people, particularly those who take the simple and direct view of garbage, who look at dogs and see man's best friend: fuzzy things hovering over slippers and newspapers and flasks of lifesaving brandy. Or, in the case of New York City, trained killers. Guard dogs and watchdogs. A gourmet garbage freak, on the other hand, looks at a dog and sees something quite different. It is usually solid and located in the gutter or the vicinity of the gutter. This is the gourmet approach to garbage.

This particular bit of gourmet garbage became the hot issue in New York City in the early 1970s. Despite war, plagues, pestilence, the rising crime rate, and a veritable VD epidemic, there was hardly a place a person could go in New York in the early morning of the 1970s and not be confronted with—to quote Jerome Kretchmer and other well-known environmentalists—"dog shit." People were either stepping in it or discussing it, walking over it or talking it over. It became the topic of political meetings, community meetings, seminars, forums. Committees were formed. Irate citizens banded together and spoke out.

"There are plenty of ugly and dangerous aspects of life in New York that are difficult to do anything about," wrote Clark Whelton in the *Village Voice*. "You may have to take a certain amount of shit from the transit system, from the phone company and from Con Ed. But dirty, dangerous dog shit on your shoes, parks and streets? That's one kind of crap you don't have to take." Whelton immediately became that newspaper's expert on New York dog shit.

There was no way for Jerome Kretchmer to escape the dog shit issue. Kretchmer, en route to a black-tie do uptown, drops in at a meeting of some four hundred Greenwich Village residents. It is the usual mixed bag of Village idiots, artists, writers, old-line Italian residents, ex-junkies, and homosexuals. Kretchmer wades in, unafraid. "The more I come to these meetings," cried an irate man, "the more I'm convinced City Hall is like Alice in Wonderland."

"Yeah," responded Kretchmer, undaunted in his floppy bow tie, ditto his mustache and tails, "and I'm the Mad Hatter."

"Whaddya gonna do about dog shit?" was the way the queston was finally put. The fight was on. Washington Square Park had become a giant dog run, one woman claimed. Another said she didn't take her dog out on the streets anymore—they were too filthy. "There ought to be plastic bags for dog owners, and dog owners would have to scoop up after their dogs." Kretchmer stood up under the whole load as it hit the fan, remarking that there was a backlog of some 70,000 sanitation summonses in court, many of them for dog shit. "We can't get the judges to pay attention," he said. "So we have a bill in the legislature to create an environmental court to clean up these cases with $100 and $200 fines."

As for banning dogs: "Politically unrealistic," said Kretchmer. There are some 500,000 resident registered dogs in New York City, each one with a potential voter on the other end of the leash. Now *that's* politics.

Dogs and dog shit are indeed a problem. Add to those 500,000 "legal" dogs an additional 250,000 illegal ones in the form of wild dogs who roam the city streets and there is really a problem. Using as a rule of thumb a pound of shit per day per dog, that's 750,000 pounds of dog shit that hits the city streets, sidewalks, parks, and property each and every day.

One of the most vehement and vocal opponents of dogs in the city is Fran Lee, former consumer affairs editor of radio station WNEW and founder of an organization called "Children Before Dogs." Fran Lee sees dogs as a health hazard and speaks out whenever she can on the subject. At one gathering, an open meeting to discuss the dangers of dog shit, an unidentified woman stood up at the back of the room, lifted her arm, and hurled a sack of shit at Fran Lee. Without breaking verbal stride, Fran Lee expertly snagged the sack and threw it back at the woman, hitting her in the side of the head as she ducked out of the door. "I think the dog crazies are getting the idea," she remarked later. "At least they're putting it in bags."

In 1971 Nutley, New Jersey, became the first city in the country to pass a law requiring dog owners to clean up after their dogs. Malverne, New York, became the second town when their "crap control" law went into effect later that year.

"The first problem with dogs is dog shit, to use the technical term," says Jerome Kretchmer on the subject. "The dog problem is a people problem. It indicates people's lack of respect for other people. When a dog is crapping all over the sidewalk, it's a person who's letting him do that, so it's an indication of how people feel about their environment.

"What's the difference between a guy whose dog is shitting on the sidewalk and a guy throwing his cigarette pack on the sidewalk? It hasn't been proved that dog shit is harmful to your health. But I can prove it's unsightly and that it doesn't smell good, and I hereby invite the next person who steps in dog shit on a Sunday morning when he goes out for the paper wearing sandals to let me know how it *feels.*"

Next to cars, what Jerome Kretchmer loves to hate most is dogs. And cats. "You got *five* cats?" he cried in disbelief to one poor reporter who confessed her animal life to him. "And a *dog*? Christ. What you need is a good man."

Not that doggie-do is the only form of gourmet garbage that the residents of New York City have to contend with. Take parades. By the end of the 1960s they had reached such a zenith that in 1969 parades alone cost New Yorkers $742,458. To walk up and down Fifth Avenue beating a drum and twirling a baton? Nearly three-quarters of a million dollars to see Irishmen and Puerto Ricans and Germans and Poles walk up and down Fifth Avenue? Surely a green stripe in front of Saint Patrick's is not that expensive.

For every parade—and in 1969 there were twenty-one major parades—it costs money to put up those reviewing stands, and bleachers for the crowds. You have to pay overtime for police since more cops are needed for crowd control.

"Unless the City acts now to assure a formal procedure for financing future parades," said a report by the Mayor's Council on the Environment, "it is entirely possible that budget cutbacks in upcoming years will force their elimination." The MCE recommended that groups post a bond, based on the clean-up costs of the previous year's parade. Parade permits would be withheld until the bond was posted. Additionally, reductions in the bond would be made if the group chose alternate days and sites, other than busy thoroughfares during peak daytime hours.

"For that same amount of money—$742,458—New York City could have paid the salaries of 900 school teachers for one year," said Rebecca White of the Mayor's Council. "There is nothing concrete to show for the parade money —no parks, no drug clinics, no permanent pedestrian malls." Not that Rebecca White is a pinch-faced old hag whose very bones clack painfully at the sound of "Stars and Stripes Forever" or "When Irish Eyes Are Smiling." Not at all. "Granted, there have been fun-filled afternoons for many; but for many others the parades have meant only inconvenience in traveling and, of course, that $742,458 bill that *everyone* has to pay."

The various ethnic parades—Saint Patrick's, Hellenic, Pulaski, Puerto Rican, Columbus—usually bring in a dona-tion of about two to four tons of gourmet garbage each. Individuals competing for the heaviest avalanche of honors are Astronaut John Glenn, whose 1962 motorcade left be-hind 3,474 tons of garbage—most of it paper—and General Douglas MacArthur, whose 1951 honors mounted up to 3,249 tons. All of this is up from Colonel Charles Lindberg's 1,750-ton total in 1927 and Howard Hughes's 1,800 tons in 1938. By far the biggest victory celebration to shovel up after was V-J Day in 1945. This celebration lasted three days, one unofficial and two official, and dumped a total of 6,334 tons of excitement on the streets of New York.

The New York City Department of Sanitation proudly claims it had one of the few accurate presidential polls during the 1948 campaign: when Truman came to town they had to pick up 63 tons of confetti, ticker tape, and miscellaneous paper. Governor Thomas Dewey, on the other hand, garnered only 9.3 tons of support from New Yorkers.

Of course there are those spur-of-the-moment parades that nobody can predict and, of course, nobody can stop. The Mets winning the pennant in 1969 brought a spontaneous flurry of every conceivable kind of paper—from toilet paper to IBM punch cards—floating out of the windows in New York City. From rest homes to the RCA Building, New York was buried under a veritable avalanche of homemade confetti as the victory celebrations began. That little burst of enthusiasm brought a total 1,254.6 tons of excitement pouring down onto the streets of New York. It cost taxpayers $9,775 in garbage bills alone to win the pennant.

In 1969 an eight-inch snowfall paralyzed New York City. It was a soft, gentle snowfall, rather wet, and there were snowmen all over the city to prove it. Eight inches of snow spread out over a period of twelve hours brought New York City to a standstill. It nearly cost Mayor John V. Lindsay the 1969 election and it factionalized and polarized a growing rift between City Hall and the Department of Sanitation—which is responsible for snow removal—that was not healed until later that year when Jerome Kretchmer blustered his way onto the scene.

Snow is always very special to New York City, simply because the city never actually gets very much and thus doesn't really know now to deal with it. People don't know how to drive in it, dress for it, react to it. When this particular snow came, the weather bureau had been rather wishy-washy in its predictions, the Sanitation Department rather lackadaisical in making any emergency plans. Even after the snow came, everybody worked at cross-purposes. The Traffic Department didn't get the automobiles banned in time, so they were stuck and stalled all over the place. The salt spreaders and snowplows didn't get out early enough. The airlines—even in the thick of the blizzard—cheerfully assured would-be passengers that indeed the flights were leaving when in fact the planes were already stuck on the runways. Hence air travelers trying to fly the friendly skies found themselves earthbound on expressways and free-ways all over the place trying to get to an airport that wasn't even functioning.

In the middle of it all stood Mayor John Lindsay and union chief John DeLury having a grand old time hurling invectives at one another. DeLury got bullish and threat-ened to refuse to let his men work overtime. The city cooperated by being mulish. Over 1,800 sanitation men who reported to garages other than their regularly assigned ga-rages (the snow was so bad they couldn't get to them) were turned away. Meanwhile the snow mounted along with the acrimony. Days later, the snow was still standing out in Queens and Mayor Lindsay found himself stuck in a grow-ing drift of resentment from irate residents who couldn't walk and couldn't drive, and whose garbage hadn't been picked up for days. They could talk, however, and it was far from polite. Then sporadic outbursts of graft and cor-ruption—rather typical, actually, for NYC—began leaking out. The sanitation men who demanded $100 fees for com-ing in and collecting garbage. Some residents of Queens, after waiting six days for city snow removal crews to dig them out, finally banded together to hire a private snow-plow at $250 a day to free them from their snowy prison.

By 1970, however, Jerome Kretchmer was on duty, super-vising the snow alert like Patton presiding over the Third Army. The snows were predicted for New Year's Eve. All day Kretchmer fidgeted and fumed, waiting for the big moment. He was down in Chinatown with Dorothy and some friends, having tucked away an enormous dinner of egg rolls and wonton and fried rice and what not. Suddenly, into the crisp night air of the New Year, the first few flakes floated down. Kretchmer looked at his watch, let out a yell. "Right on!" he cried, referring to the fact that the snow predictions were right on the button.

And the snows came.

When a salt spreader passed Kretchmer and his merry band down in Chinatown, Kretchmer jumped on the back for a brief whirl through Chinatown. Then he stayed up all night deploying his forces through snow and gloom of night. When dawn came Kretchmer was still at it. He topped the whole performance off with a television appearance telling everybody how smoothly things were running.

It cost the city about $1 million to cope with that mere eight inches of snow. (Parts of Maine are buried under fifteen feet throughout the winter. Snow falls in eight-inch hunks nearly every week out in the Great Plains.) And feisty old John DeLury had a new hero as he applauded Jerome Kretchmer as the first sanitation chief he'd seen in his thirty-six years as union chief going "out into the battlefield" while the fight was at its height.

The United States Government—particularly the Defense Department—wins the gourmet garbage prize hands down. The list of exotic DODrop-outs is impressive in excess: the XB70 bomber; the supersonic transport; the Air Force's manned orbiting laboratory; the nuclear-powered airplane. Each of those projects had price tags averaging about $1 billion—*each*. After thirty years of running hot and cold Wars, the U.S. Department of Defense has achieved what many thought to be the impossible. It is indeed the Garbage Gartantuan. Every time the DOD changes specifications, for example, or decides to scrap costly and complicated weapons systems, the litter piles up. *Time* magazine has even accused the U.S. Department of Defense of becoming "far and away the nation's biggest litterbug."

When we were building our DEW line (Distant Early Warning) radar system in the Arctic, the DOD discarded some 100,000 oil drums on the shores of the Beaufort Sea, within the boundaries of the nation's largest wildlife refuge. Some were only partially emptied by the departing military and are now leaking oil, which is, of course, highly toxic to the wildlife on whose homes they are resting.

Camp Kilmer in New Jersey is a veritable ghost town of decaying, fire-gutted barracks, and shattered glass. California's Anza-Borrego Desert State Park, all 10,000 acres of it, is off-limits to visitors because it still contains unexploded bombs and rockets, donated by the U.S. Navy thirty years ago when the park was used as a test-firing range. Although much of the bomb garbage is buried beneath the sand—or nestled in the bomb craters that pock the park's surface— a civilian salvage team that went in searching for scrap five years ago suddenly disappeared in a cloud of smoke.

Firmly believing we should carry our garbage know-how to the outer reaches of our empire, the DOD has done an excellent job of doing just that. Alaska and the Aleutian Island chain are littered with a potpourri of military garbage. More than 2,000 World War II quonset huts blister the landscape at Amchitka in the Aleutians. Bomber tails and ruptured fuselages are scattered about. It is estimated that over one million fuel drums are lying around and about the north coast of Alaska.

Obviously not satisfied with the quality of the gourmet garbage on Amchitka—it is, after all, old and out-of-date— President Nixon was firm in his commitment to do something about it. Thus, with callous aforethought, he ordered up the now legendary "Amchitka Test" back in 1971. It was a dramatic example of twentieth-century garbage generation, for the test was a marvel of American technological know-how and military might. Buried 6,000 feet down in a cave on Amchitka Island was a hydrogen bomb that weighed in at five megatons (translation: the equivalent of

about 5 million tons of TNT). The purpose of Amchitka
was to test a nuclear warhead for the proposed Spartan
defense system, which, theoretically, seeks out ballistic
missiles and shoots them down. All this activity took un-
told years of research and development and design work,
plus four years of actual preparation, at a cost of about $200
million. It was over in one ten-millionth of a second. And
down below in that cave? An awful lot of garbage left over.
Melted metal and plastic and wires and smashed atoms. Up
above, cracks in the earth, dust clouds. No sooner was the
blast over than America announced her plans to vacate
Amchitka with all due haste, leaving behind silos, bunkers,
command posts, and that big, hot hole in the ground.
Landfill—now radioactive—in the Western Aleutians.

There were protests and pickets and parades from the
United States to Canada to Alaska. But once the green light
was given, in typical all-American fashion it was A-OK all
the way. One truly outspoken critic was Dr. Wolfgang K.
H. Panofsky, director of the Stanford Linear Accelerator
Center at Palo Alto, California. "The test has no real military purpose," he said, "and is being held because of technological inertia." His was the whole depressing technological argument summed up in grim, pragmatic terms. As
far as Panofsky was concerned, "The issue is whether man
controls technology or technology controls man. The people who design bombs just don't want to stop. The question
is whether to continue to test just because we've done the
work leading up to the test."

Well, the test went off with a bang! that was felt around
the world. So far there have been no earthquakes or tidal
waves or other freaky environmental disasters. But perhaps
that was not the point. Perhaps the point was even worse:
that once again technology won. That once more a certain
portion of our priorities—years and years and years of
them—would be devoted not to education or housing or

prison reform or food for our hungry children. That once again it was technology, not society, that was the winner. That the results of Amchitka were more bad vibrations and more solid waste.

If the Army can generate the likes of millions of oil barrels in Alaska when there isn't even a war on, think what they can do when confronted with the real thing. Make no mistake about it, the business of war-oriented solid waste is big business indeed. The biggest business we have. World War II is estimated to eventually cost U.S. taxpayers a total of $664,000,000,000. Original war costs— shells and bombs and khaki socks and K.P. rations—were $288 billion. Add to that G.I. benefits, war loans, and the like, and the cost skyrockets. Vietnam, the next most costly conflict, will eventually hit $352 billion. America stands to have invested $1,311,907,000,000 in wars since the Mexican– American number of 1846. Again, in round numbers, that is 1 trillion, 311 billion, 907 million dollars. Or, in even simpler terms: a lot of money.

But that's all ancient history. World War II is merely a passage in the history books compared to today's war efforts. A baby boom compared to the blast that the DOD is determined will constantly confront us.

Vietnam is the second biggest war we have ever had: 45,933 Americans killed.

Vietnam is the longest war: it has been twelve years since the first GI was killed back in 1961.

Vietnam is the most productive war we have ever had: a report issued by the Senate Subcommittee on Refugees, chaired by Senator Edward Kennedy, stated that in addi- tion to those 45,933 dead American soldiers and the 303,616 wounded there had been 1 million South Vietnam casual- ties, both military and civilian. The war has created 8 mil- lion South Vietnamese refugees (almost one-half the popu- lation), 1 million Laotian refugees, and 1.5 million

Cambodian refugees. In 1969 alone some 25,000 South Vietnamese civilians died in the conflict and over 100,000 were wounded.

Vietnam has been one of our most costly wars. Economist Robert Leachman put the original war costs at the end of 1972 at $400 billion and that, he said, "is a conservative estimate."

Back home, one of the more sophisticated forms of gourmet garbage is the old, outmoded chemical or biological weapon: old cannisters of nerve gas; antiquated strains of bubonic plague. What to do with this stuff? Rail-hauling it across the country for ocean disposal usually causes a great hue and cry from the communities this triumphant processional is set to pass through. Underground burial is not eyed too keenly by those residents of the surrounding area. Obviously cremation is out—air pollution and air-borne particulates.

What had to be done was to develop an entirely new disposal technique for the biological weapons we had been stockpiling from 1953 until 1969, when biological weapons were supposedly banned. We were stuck with:

Pasteurella tularensis (commonly known as rabbit fever)

Bacillus anthracis (responsible for anthrax, usually fatal)

Venezuelan equine encephalomyelitis

Coxiella burnetii (the Rickettsia responsible for Q fever)

Botulinum Toxin A (a dangerous food poisoning, very much like the Bon Vivant Vichyssoise variety)

Staphylococcus enterotoxin (another type of food poisoning)

"One of the benefits of biological and chemical warfare," said Colonel John Stoner, commander of the Pine Bluff Arsenal, where one of the nation's largest stockpiles of biological agents was created and kept, "is that you can incapacitate people without killing them. It's more humane."

Humane or no, after 1969 they had to go. It was quite a production: inside a windowless, ten-story brick building, workmen wearing special white coveralls and gloves were handling trays of germs. The germs first had to be transferred to separate holding tanks where they were sterilized with steam heat to 280° F. Then they were cooled and tested to make sure the agents had been destroyed. Then the inert remains were treated very much like domestic sewage, sterilized again at 280° F., tested again, and then moved out to a nearby sewage treatment plant where they were further biodegraded. Once that was done, the hopefully dead germs were dried, collected, spread over arsenal property, then disked into the soil to a depth of four inches and planted with grass.

Just to dispose of this particular bit of garbage cost $10 million. Not that that $10 million got rid of it all, either. It was reported in the October 11, 1972, *Congressional Record* that "the Army still maintains unspecified quantities of disease-producing germs 'for research purposes' (including the deadly Anthrax). And the Army has not dismantled its large germ-producing facilities."

So much for biological agents. But there still remains a colorful array of *chemical* weapons: old mustard gas and nerve gas. The Army had 2,071 tons of obsolete WW II mustard gas stored in steel canisters in its Rocky Mountain Arsenal outside Denver. First it was going to rail-haul them across the country and dump them in the ocean. Cries of alarm and protest echoed along the route of the train, and finally the Army capitulated and canceled the order. Final disposition consisted of burning the canisters of poison gas at an incinerator ten miles from downtown Denver. (The Army promised not to burn on days when pollution was high to begin with, since even mustard gas puts a lot of sulfur dioxide and hydrogen chloride into the air.)

But all of these—dogs and parades and snow and individual little atomic blasts and decaying WAC barracks—these are all small fish in the larger ocean of garbage we are now producing. Hardly worth a mention, much less our attention. And certainly not worthy of anything approaching awe or appreciation. Not when compared to nuclear wastes, that is.

The discards of our nuclear arms race through the seemingly infinite space of the cold war. There is, in Hanford, Washington, the world's biggest collection of radioactive wastes in the world at the Hanford Atomic Research and Production Plant. There it sits, ticking away while scientists and politicians and the Pentagon and the AEC argue about who's responsible and, more importantly, just what can possibly be done with it. Normally, radioactive wastes are stored in liquid form in tanks. Unfortunately, due to the high corrosion factor of atomic wastes, these tanks usually have a life-span of about twenty years. Time is running out at Hanford.

One projected plan was to load the lethal discards onto rockets and fire them off to the heart of the sun, where they would spent the next few million years decaying and thus becoming harmless. This was scrapped both because it was (1) too expensive and (2) unsound because of technological problems. (We can get a man to the moon, but not junk to the sun.) Then someone had the bright idea of burying the nuclear junk in all those abandoned salt mines out in Kansas. Plans were made, projects drawn up. Then the people of Kansas objected. Who wants radioactive garbage, they reasoned. The AEC had it all worked out: dig holes in the bottom of the salt mine—a cavern roughly the size of a football field—then stick canisters containing radioactive garbage into these holes, then fill the cavern back up with salt. But here's where it gets a bit tricky: after a very short time—six months to ten years—the steel canisters surrounding the radioactive wastes will have melted away with corrosion, thus leaving the radioactive wastes encased in salt, which, officials are quick to explain, has a plastic consistency like candle wax under these conditions.

The public outcry could be heard all the way to the corridors of the AEC bureaucracy. Eventually plans for the atomic graves were scrapped and now, according to Kansas Representative Joe Skubitz, "the Kansas site is dead as a dodo for waste burial."

So the AEC keeps working. They are now thinking of burying them in old volcanic lava beds right under their facility at Hanford.

The list of radioactive garbage is endless. Think about those people out in Grand Junction, Colorado, who built their homes on radioactive wastes left over from mining uranium. Nobody wants it. The government—who wanted the uranium in the first place—doesn't want it. The mining company, with typical élan, just walked away from it once the mining was through. Looking and acting very much like sand, the stuff was then used in fill and foundations for homes in Grand Junction and environs. Playrooms and laundry rooms and living rooms are all built right on top of radioactive sand. A study done by the U.S. Department of Health showed increased birth defects, higher cancer rates, and lower birth rates in the affected area compared to the rest of the state. Cleft lips and cleft palates, for example, were double the usual rate. Doctors participating in the study found cancer deaths "significantly higher."

Meanwhile, the buck gets passed faster than a speeding bullet, with nobody wanting to take the responsibility—because then they might be stuck with the awesome task of doing something about all those radioactive homes. And that costs money.

The Grand Junction area is not the only area with radioactive tailings. Huge hills of tailings can be seen all over the Southwest, familiar landmarks in Colorado, Wyoming, Texas, Arizona, New Mexico, and Utah, as well as South Dakota, Washington, and Oregon.

The defense of the motherland had some serious competition, however, during the hi-ho years of space exploration. In 1965 the NASA (National Aeronautics and Space Administration) budget was $4,562,000,000. The next year it was up to a high of $5.4 billion. Although it has been steadily declining since then, the NASA budget for 1970 was still nearly $4 billion. As Jerome Kretchmer, head of New York City's Environmental Protection Administration, says, "Yeah, terrific—we put men on the moon. And they leave behind a few million dollars' worth of solid waste."

The celestial garbage—worth $517.2 million—generated by NASA is certainly something to ponder, for this is indeed the stuff dreams are made of. Think of those six plastic flags standing proudly at attention up on the moon, forever unfurled and silently unflappable in the eerie quiescence of the moon, held at stiff attention through time and space by hidden wires. Think of the three moon rovers (at a cost of $6 million each) and six television cameras decorating the lunar landscape. The camera equipment alone is worth $5 million. Think of the tire tracks forever and indelibly etched in the stillness of the moon's surface, left there as solid-waste reminders of the American presence. Reminders of that one big step for mankind.

And that's not all. A compilation of the garbage generated by lunar landings alone is both impressive and extensive: it runs to fifteen pages and contains over 344 entries. Along with those six flags and six television cameras and three lunar rovers there are at least eight pair of overshoes and boots, earplugs, hammocks, towels, scientific testing equipment, hammers, helmets, and some twenty-five still cameras. Think about Alan Shepard's two golf balls, Dave Scott's feather, the two Yo-Yos from Apollo 14. To get the moonmen back to the mother ship, they are blasted off the moon by the lunar module, which is then sent crashing back onto the moon's surface to test seismographic impact. There are now six lunar modules—valued at $270 million —up on the moon. Plus $130 million worth of scientific packages and $600,000 worth of tools.

Orbiting around the universe as of April, 1972, were some 2,767 man-made space objects. This out of a total launch of 5,937 since Sputnik I back in 1957. "Some objects now in orbit will be up there in the void of space—where there is little atmospheric drag—for hundreds, even thousands of years," say the NORAD (North American Air Defense Command) orbital analysts whose job it is to track these celestial cookies. Even more depressing, on the "decay-prediction" log at NORAD, are more than a score of satellites with a life-span calculated in excess of 10,000 years. And we worry about smog cutting off the sunshine.

We are right now in the midst of creating the first space-age ghost towns. A $1.4 billion effort at Jackass Flats in Nevada was the first. This fifteen-year project was NERVA, Nuclear Engine for Rocket Vehicle Application. It began in 1957, one of the first giddy new space projects. Year after year appropriations were blindly approved by Congress. "We've spent $1.4 billion," Senator John O. Pastore (Democrat of Rhode Island), chairman of the Congressional Joint Committee on Atomic Energy, cried angrily last year. "The big question is *why* did we spent this $1.4 billion?" That $1.4 billion lies scattered and abandoned across the hundred square miles NERVA appropriated in the Nevada desert below Skull Mountain near, appropriately enough, Death Valley. Some $200 million worth of deserted engine test stands, gigantic assembly bays with birds nesting in the rafters. Empty administration buildings. The aluminum and steel remainders of pumps, turbines, and motors. In its apogee, NERVA commanded $100 million a year, no questions asked. As Senator Pastore said, "Why?"

This is not to say that garbage has to be something vast and expensive and technologically beyond the reach of the everyday, ordinary tax-paying American citizen. In addition to the usual run of garbage we produce every day, there are moments of elitism even within that egalitarian pile of garbage.

Christmas, for example, is something just about everyone participates in. Roger Lovin writing in *The Mother Earth News* estimates that some 500 million Christmas trees are sold each year. "They die," he says morosely. "They die for the sole purpose of bringing into dreary little boxes the same sight and smell the inhabitants could get free every day, simply by seeking out the land they have fled with their bodies and yearn for with their souls."

Nowhere has the idea of garbage been carried to such heights—depths?—as in the case of one A. J. Weberman. A.J. operates on the theory that you are what you throw away, and with this in mind he has made himself one of the world's leading experts on garbage-can contents. Indeed, A.J. may be the world's *only* garbologist, a singular recognition in this world of ever-diminishing individuality. For some incredible reason known not so much by God as by his fellowman, A.J. rummages through Bob Dylan's garbage.

A.J. is something of a visionary. He can see in objects things other men cannot. Hence his passion for garbage. It began with his intense dedication to things Dylan. A.J. would spend hours listening to Dylan's music, ferreting out hidden meanings and secret intentions. They were so hidden, in fact, they were a secret from even Dylan himself. Undaunted, A.J. kept working, collecting everything Dyl-

an ever sang or said, or anything ever sung or said about
Dylan. Then—it happened. "It all started one Saturday
night when me and my old lady, Ann, were walking past
Bob Dylan's Greenwich Village brownstone,"[1] explains
A.J. "I recalled my visit to 'the king of folk rock' the week
before. That dude had some nerve throwing me out! He
knew I'd been studying his poetry for years, that I had
more insight into it than any other rock critic. Did Johnson
throw Boswell out? What a lot of garbage." *Garbage!* The
light bulb went on. "I reached into Dylan's plastic can,"
A.J. recalls, "and pulled out a crumpled sheet of paper. It
was a draft of a note to Johnny Cash." Right then and there,
A.J. knew that that "ain't no garbage pail, it's a gold mine."

From then on, A.J. was not only dedicated, but relent-
less. He went through Dylan's garbage can like a chain
smoker through a pack of 100-millimeter longs. "I was
hooked." Dylan, the Howard Hughes of rock, even hired
a garbage guard to protect his discards from the dedicated
ferreting of A. J. Weberman.

In going through Dylan's garbage A.J. learned that not
only did Dylan and his old lady have five kids—three of
them not entirely toilet-trained—but they had a cat and a
dog they were in the process of trying to housebreak. Weberman valiantly waded through a sea of Pampers (used)
and soggy newspapers (ditto used). Undaunted, Weberman persisted. One haul included such momentos as
Kitty Litter, the packaging from a box of Evenflo nipples,
assorted packaging from take-out orders from a corner
Blimpie Base, a torn copy of the rock magazine *Craw-
daddy*, plus many Morton frozen pie packages and a
sketch of the late Jimi Hendrix. Perhaps the best of all
was a shopping list done on yellow ruled legal paper.

---

1. On MacDougal Street near Bleecker, crossroads of a thousand lives and right
near the Blimpie Base fast-food parlor.

Sour cream, sugar, vannilla [sic], walnuts, swordfish, baking powder.

A.J. was hooked. The sky is the limit, he figured, this self-made Horatio Alger of the garbage. There was indeed nowhere to go but up. Uptown, that is. And so it came to pass that A.J. expanded his garbage horizons. He trekked up to the chic East Side, determined to rip off socialite Gloria Vanderbilt Cooper's garbage.

He found it locked up, behind bars.

Undaunted, he haunted the East Side. "On my first visit to Neil Simon's pail early one Monday A.M. I found the leftovers of what is for some dudes a typical New York City Sunday breakfast: a practically untouched whitefish, a half-eaten bagel, scraps of lox, and sections of the Sunday *Times.*" A.J. admits debating about sharing the whitefish with Simon but he "decided it was unscientific to eat my findings."

Back downtown a quick rummage through Yippie leader Abbie Hoffman's garbage showed *Time* and *Ramparts* establishment cheek by radical jowl. When A.J. ventured out to Muhammad Ali's New Jersey home, the butler *gave* him the garbage, which included an empty can of Luck's collard greens and one of Sunshine black-eyed peas, plus the label from Shabazz bean pie, along with an empty roach bomb and an empty pack of R. J. Reynolds cigarette papers. A.J. was not impressed. "Ali's garbage is sparse. Needs dirtying up if he wants to make it in my book."

Now A.J. is not content to rest on his research. Already he is at work on a book entitled, appropriately, *You Are What You Throw Away.* As A.J. says, "Remember: garbage is powerful."

A.J. might be able to rummage through New York garbage to his heart's content—provided the cops and/or the donors do not object—but a California court recently ruled that garbage is a very private matter. When L.A. cops went through the garbage of one Edward Krivda they discovered marijuana debris. That was enough for L.A.'s finest. They arrested Mr. and Mrs. Krivda, who moved to have the evidence suppressed. The state supreme court agreed, saying that "the question remains whether defendants herein had a reasonable expectation of privacy with respect to the contents of the trash barrels." The court figured the Krivdas had a "reasonable expectation that their trash would not be rummaged through and picked over by police officers acting without a warrant."

Lest the feminists across the land raise a clenched fist along with a hue and cry over discrimination, A.J. Weberman has his female counterpart in Pat Eakins, a free-lance writer and amateur garbologist in New York City. Pat is an observant woman who, while wading through the urban junk pile that is New York City, suddenly figured out that, like Gaul, *garbage est omnis divisa* into three parts. It is the greening of garbage. First there is the Consciousness I junkie, "at the top of the dung heap, engaged in what sociologists call 'creaming.'" To Ms. Eakins these are "relatively rich aesthetes" who buy Victorian beaded purses in expensive antique boutiques. "C-I junkies are dilettantes," Pat sniffs. Next down the line in the garbage can is the C-II junkie, who browses through secondhand stores and Salvation Army stores and the "Buy Lines" columns of the daily newspapers. Pat figures Superman was probably a C-II junkie. "Either that, or Clark Kent shopped at Robert Hall." Either way you cut it, "the C-II is a closet junkie." He buys secondhand but hopes it looks first-rate.

Now, the real nitty-gritty of Pat's theory is the C-III junkie. He takes what he can get, aesthetics be damned. The C-III junkie is into survival. Beggars and younger sisters are typical of the C-III junkie, with little or no choice and/or control of the junk in their lives or the reasons for it. "Well within the C-III ranks are hoboes, bums, winos, artists, madmen, saints, gypsies, clowns, paupers, and magicians," Pat figures. "Robinson Crusoe was a C-III junkie. The willing C-III junkie is the prototypical junkie, and all others are mere imitations.

"Since the natural habitat of the C-III junkie is the trash can and the gutter," says Ms. Eakins, herself a confirmed C-III junkie, "don't imagine this milieu makes him good-tempered." She speaks from experience: "A couple of hassles with a sanitation engineer who covets what the junkie has extricated from the coffee grounds, grease, paint rags, used Kleenex, and stinking Kitty Litter makes him secretive, surly, and paranoid."

Pat sees the archaeologist as a good example of a high-class, C-III trash picker. The effete snob of the garbage can. "They aren't ashamed to poke around in it," she says. It's just a difference in attitude toward the archaeologist who digs garbage. "The same smug swells who snicker at me as I go through the junk in New York garbage pails 'ooh' and 'aah' over all the potsherds the archaeologists pick out of petrified middens in ancient dumps. 'Archaeologist' is a euphemism"—she sniffs—"like 'sanitation engineers.'" What archaeologists have done, she figures, is give junking some much-needed redeeming social value. As Ms. Eakins grimly and glumly sees it, "Until Lee Radziwill is proud to be seen in Jackie Onassis's hand-me-downs and Pat Nixon and Martha Mitchell are photographed going through each other's garbabe piece by piece, ecology will not be a watch-word and waste will be a problem in the land."

But if America is anything, she is still the land of hope. "This ecological millenium can be hastened," Pat figures. "Give a piece of trash a good American home and give a shot in the arm to a nodding civilization." But what if folks object to the idea of being an urban junkie? Easy, says Pat. "Call yourself an archaeologist. Or a midden maven."

Another much ignored area for gourmet garbage is in our hospitals. Hospital garbage is generally classified in two categories: (1) general, the kind generated by most large institutions such as hotels; (2) contaminated, that refuse composed of surgical specimens, dressings, amputations, and the like.

Hospitals, like almost every other place in our society, do one of two things with their garbage: burn it in on-site incinerators or cart it away where it is either burned or buried. Needles, gowns, arms, legs, after-births, old discarded appendices, wornout hearts, unwanted tonsils. The whole works.

Despite the arms and legs and hearts and tonsils (which are generally incinerated on the premises) the biggest hospital headache is disposables. For one thing, disposables are cheaper, especially by the dozen. A dozen standard-size, multiple-use syringes cost about $23. One hundred disposable units cost about $6.50. (Often the disposability of such things as syringes is a myth. In New York City, for example, the disposable syringe is effectively recycled back into the community by junkies who rummage through hospital garbage and pluck out the discarded syringes.)

There are today 7,000 general hospitals treating some 31 million patients each year. Ten years ago each of those patients generated about five to ten pounds of garbage during that hospital stay. Today each hospital patient contributes fifteen to eighteen pounds of garbage—most of it disposables: presterilized bed pads, sheets, blankets, examination gowns, diapers, washcloths, shoe covers, curtains, thermometers, syringes, water carafes, drinking cups. There are 2.5 billion meals served in hospitals each year, many of them on plastic, disposable, throwaway plates and trays.

One hospital administrator found that the straw that broke the back of his disposable camel was the single-use patient service tray (individual to each patient) with plastic water pitcher, plastic cups, and plastic thermometer. The patients, instead of taking them home as the hospital counted on their doing, just left them to be thrown away by the hospital. The result was more bulky solid waste for that particular 700-bed hospital. It finally had to invest in its own garbage truck armed with a compactor and then truck its own refuse to the dump. Edward Bertz of the American Hospital Association would like to see hospitals be absolutely discriminating in their use of disposables, using germ control as the primary criterion rather than disposables for disposability's sake.

The New York City EPA called the exploding use of disposables in hospitals the fastest growing waste disposal problem. In 1971 the Christ Hospital of Cincinnati called for a moratorium on the further use of disposable items until the hospital standards committee could figure out a set of priorities for their use. An editorial in the *Journal of the American Medical Association* urged other hospitals to do likewise. "The person who is responsible for bringing in disposables," said one Long Island hospital administrator, "is not responsible for getting rid of them."

"Disposables have become a part of the American way of life," says Sister M. Diana Salkowskik, who edited the report of the Second National Conference on single-use items in health care. "They are timesaving, attractive, and convenient." She says it's nice that hospitals provide those cute take-home disposables for patients. But what about the mother who sends four kids in at once to have their tonsils removed? What about the patient who comes in regularly for treatment of a chronic ailment? "Recently, one mother when told she could take the thermometer and utensil kit home remarked, 'Why should I? We have over a dozen of each at home now.' Today is the time for solving yesterday's problems created by the use of disposables and the time for preventing new ones from rising tomorrow, particularly in the field of waste management," summed up Sister Salkowskik.

Again, it is just a simple matter of priorities. No one would think of replacing disposable tubing with reusable rubber tubing, highly contaminated and contaminating. Disposable syringes make sense. But why all those disposable food trays and plates and knives and forks for patients who are not contagious and contaminated? Edward Bertz isn't even too sold on the idea of disposables as a way of keeping germs down in and around the hospitals. He would like a wide-ranging study on disposables, which would, for one thing, figure out just how germ-free a disposable product was after having sat on a shelf for several weeks or months.

Think about disposables. Cigarette lighters whose function is proudly announced to be their disposability. Whole generations of kids wrapped in the plastic swaddling of Chux and Pampers. Thomas Lane, vice-president and general manager of Johnson & Johnson Company, figures there are about 25 *billion* diaper changes in the United States every year. "We are dealing with a potential problem," he says, because "it has the capability of becoming a billion-dollar-a-year market for just one disposable." Breasts and bottles are a thing of the past, except to Princess Grace and a few other *La Lèche* followers. Now Mommies can use "Toss 'ems," which are just as the name implies: plastic inserts, for glass or plastic bottles, which, when used, can be tossed away. (Naturally they come in their own handy plastic dispenser pack.)

The list is depressingly endless. From paper napkins to disposable toilet mops that, when used, the consumer just flushes down the toilet (unfortunately into what is undoubtedly an already overburdened municipal sewer system). Scouring pads that are billed as just dandy for cleaning up one meal, whereupon the user is encouraged to throw them away. In 1972 Maxwell House introduced their new "Max-Pax," which encased enough ground coffee to make a potful of coffee in a plastic pak that, after use, could just be lifted out of the pot and thrown away. God forbid the Maxwell Housewife would want more or less coffee than Max-Pax provided for.

But perhaps our disposable way of life—our leaning toward plastic perfection—reached its most depressing heights out in southern California. Southern California, which used to spell year-round greenery until the automobile and ribbons of concrete and patchworks of tract homes came along and real ones couldn't or wouldn't grow anymore. Early in 1972 Los Angeles planted the median strip of an expressway with plastic trees and shrubs.

EPILOGUE

There was a time back in the good old days, when I would go into paroxysms of joy at the sight of a Woolworth's. Entire carefree days could be spent strolling the aisles of a Safeway Store. The Big Apple meant not so much my arrival in New York as a new kind of supermarket. Department stores enthralled me. Mine was an innocence bounded on the one side by a dime store, on the other by a supermarket. Happiness was a shopping list. Periodically, just for kicks, I would toss away my list and indulge in a little madcap impulse buying.

Then one day I had my consumer consciousness raised to dizzying heights. No longer could I carelessly stroll through a dime store or a supermarket. No. The Dark Age of Enlightenment was upon me.

To walk into a store meant to walk into an exhibition of excess packaging. I was awash in a sea of polyethylene, polystyrene, polyvinyl chloride. I was strangled by Saran

Wrap, assaulted by convenience foods, pumped full of preservatives and additives and chemicals. I became aware that
we were not so much sitting down to a meal as engaging
in some sanctioned shooting up at our own dinner tables.

I had to take action. I could not sit idly by while the
world buried itself in garbage and slowly committed suicide with a lethal blend of food additives and flavor enhan-
cers. An article told me that rats, when fed a diet of commercial white bread, either died or cannibalized themselves. Good God, I thought. Garbage food. My neighbors
aren't safe from me. I might run next door and nibble a bite
out of John Church's toe any day now. I read on. It was
worse. In bleaching the flour for commercial bread baking,
76 percent of the iron is removed, 89 percent of the cobalt,
and 78 percent of the zinc. The body uses iron for oxygen,
cobalt for building up healthy red blood cells, and zinc for
such things as healing and dwarf prevention. I plunged
elbow-deep into my mixing bowl.

It was hard. First, I realized that all my supermarket
stocked was the same old cannibalistic flour my plastic
bread was baked with. Not to mention the fact that the
grains were grown with chemical fertilizer, were sprayed
with pesticides, and had all those preservatives and additives pumped in. I headed for the local health food store
where I paid a highly inflated price for organic flour that
nobody had bothered taking anything out of so they could
put something back into. That was Wednesday and, being
something of a sentimentalist and a stickler for tradition,
I had to wait until Saturday to do my baking. (Saturday was
always baking day at my house.) I nearly cannibalized the
dog, I was so hungry for a piece of toast.

Saturday finally arrived. A neighbor dropped in to
watch. The whole day was shot waiting for the yeast to get
working, waiting for the bread to rise, pummeling it back
into submission so it could rise again, like phoenix out of

the ashes. Finally, it was in the oven. At last—a month's supply of beautiful bread. It was so beautiful I sat down and cried.

I was exhilarated, headed for the good life, rid of excess packaging and convenience foods and additives and preservatives. No more plastic bread bags to clutter up my garbage can. Instead, a healthy mix of iron and zinc and cobalt clang-clang-clanging away in the trolley of my stomach.

I threw out all the canned cat food, thinking of all those cans I would be conserving along with all those protein-starved Peruvians and all the endangered whales that were being sacrificed so my five stray cats could eat. I began cooking liver and kidneys. But wait: fertilizers, pesticides. I went to dry cat food. Oh, no. Additives and preservatives. I settled on dry food as an interim solution. It was disastrous: Tom Cat had to have his penis amputated: dry food gums up the kidneys or some such thing. By this time the family basset hound was in a decline because in my attempt to save the wild mustang I had switched to dry dog food. Her skin broke out and she got hooked on morphine as the treatments dragged on. (The next year I tried to cheat on my income tax. Fortunately it was my tax man and not the IRS man who intoned those fateful words to me as he waved my medical deductions under my nose: "What is this, Miss Kelly? Three hysterectomies, two castrations, and a penis removal?")

I tried not to think about it. To distract myself I switched to white toilet paper because I had read that the dye in colored toilet paper mucked up our streams and rivers. Then I read it didn't make any difference, but by that time I was too embarrassed to switch back to color-coordinated toilet paper because by now that seemed frivolous.

But worse, my whole life-style was ruined. No longer could I take a walk down the aisle of my local supermarket or my neighborhood dime store. No longer could I even go past the corner newstand without seeing it in terms of 50 percent paper in our solid waste stream. A supermarket began to equal junk food and excess packaging. The dime store turned into a massive monument to disposable (breakable is more like it) plastic toys and synthetic stretch pants.

I went cold turkey from paper napkins, switched to cloth ones. To save our diminishing fresh-water supplies, I made sure I was buying eco-approved laundry detergent in which to wash the cloth napkins. I bundled my papers for the neighborhood recycling center, then got a summons from the fire marshal for keeping them stacked in the hallway, thus creating a fire hazard. Then a neighbor and I worked out a share-the-*Times* routine whereby I read it early in the morning and he read it when he got up at noon and tried to piece together the day's news around the clippings I had made. I stomped the few remaining tin cans I would allow into the house as if it were 1943 and Hitler were right outside my front door.

I wrote endless letters (on recycled paper, of course) complaining about excess packaging and convenience foods and junk foods. Then I got a guilty conscience about all that paper that was flying back and forth (presidents of major corporations do not, as a rule, use recycled paper) and stopped. I began doing my bachelor dishes only once a day to conserve water and diminish the supply of detergent entering our water supply. Naturally the cockroaches discovered this and invaded me, which, of course, could only be halted by spraying them with pesticides.

I stayed away from fish to avoid mercury poisoning, meat because it was just too damned expensive.

I tried to dovetail my war contractor's list from Another Mother for Peace with my ecological interests and soon found I would probably be starving to death while wearing wrinkled clothes as I desperately tried to avoid General

Electric, General Motors, Honeywell, and Dow products and any grant from the Ford Foundation (which holds stock in the Honeywell Corp.). [1]

Then I thought about all those people I was shoving out of work because I wasn't buying all their plastic and paper and aluminum and their fish and fowl and foul products. Good heavens. Visions of large-eyed, hungry children in shabby clothing forced to leave their $30,000 suburban home, sell the second car, and hit the expressway in search of the nearest welfare office assaulted me. It was almost more than I could bear.

Indeed, life is but a series of compromises.

But beyond cat and dog food and baking my own bread and giving shoppers dirty looks when they reached for junk foods and excessively packaged fake foods, I felt more positive action was required. It was one thing to give dirty looks, it was another thing to take a public stand. And so it came to pass that I determined to rip off my excess packaging and leave it at the check-out counter.

When the big moment came I was struck by something resembling what I felt when I had to recite "A Visit from St. Nicholas" at the fifth grade Christmas program back in Albion, Nebraska. Stage fright. I was terrified. I'll come back later, I thought, making a 360-degree turn away from the check-out counter. I wheeled that cart around the store for the next one and a half hours. The off-duty cop moonlighting as an assistant manager began giving me dirty looks, as if I were engaging in a new form of hustling. Finally, I could stand it no longer. I took a deep breath and pulled into the checkout counter. I kept my eyes shut, reached for the eggplant, and with a defiant gesture I ripped off the plastic wrap and liberated the eggplant from its plastic prison. I had struck a blow for freedom!

---

1. Honeywell makes "guava bombs" that pierce only flesh.

"You, too?" the girl at the check-out counter noted.

I opened my eyes and proceeded to free all my vegetables and take the shrink-wrap and cardboard holders off the razor blades and flashlight batteries. Check to make sure I had returnable bottles around my beer. Recycled cardboard egg cartons instead of polystyrene. I refused double-bagging, haughtily turned down a plastic shopping bag. It sure is harder to go shopping these days. I noted with a sort of humble pride the pile of packaging I had left behind. Packaging I didn't have to lug home and clutter up my garbage can with. Packaging my garbage men didn't have to haul away. (It's a sort of spread-the-wealth, though: some garbage man had to pick it up and put it down).

Oh, lord, ain't it hard.

You bet, Sister.

Not that I, or anyone for that matter, really want to revert completely back into the good old days. I don't want to beat my carpets or dash my laundry, along with my hopes, on the oily rocks of our polluted shores. Nor would I for one minute give up motorized transportation for walking or riding a horse. I will, however, think yearningly about quick, quiet, rapid mass transit. I will ride my bicycle when I can instead of buying a car or taking a cab. I will look for just such painful and pitiful few alternatives as are grudgingly given us in this our overtechnological, overpolluted, overpopulated, overgrown society.

What I want are simple things. I want detergent that doesn't pollute, toilets that flush with a minimum amount of water. I want the Big Three in Detroit to think less about the profit motive (according to Ralph Nader, GM grosses $2.5 million *per hour*) and more about making (1) safe automobiles that (2) don't pollute our air. Because, in the end, we all pay for it—stockholders and taxpayers alike.

I don't want to have to spend my time ripping off excess packaging in my supermarket. I would rather be planting tulips or volunteering my time to stuff campaign literature (printed on recycled paper, of course) into envelopes. Why, we ask with all the simplistic foolishness of people who must eat to stay alive, must the iron and zinc and cobalt be removed from our food so that chemical substitutions may be made?

I felt pretty good until I got home from the supermarket and started unpacking my liberated foodstuffs. The steak turned into a cow replete with massive doses of hormones, preservatives, and food dye to make it look rich and red. The eggs filled me with visions of arsenic (put into the chicken feed to make for bigger and stronger eggs). The veal cutlet, instead of manifesting itself as veal scallopini at a successful dinner party, turned into a calf with warm brown eyes.

Christ, I thought. It never ends. I can't even become a vegetarian: just the day before I had read an article about the pain threshold in fruits and vegetables. The thought of an artichoke squawking in pain every time I plunged it into the hot butter sauce filled me with dread.

Maybe I'll commit suicide with an overdose of Dream Whip.

INDEX

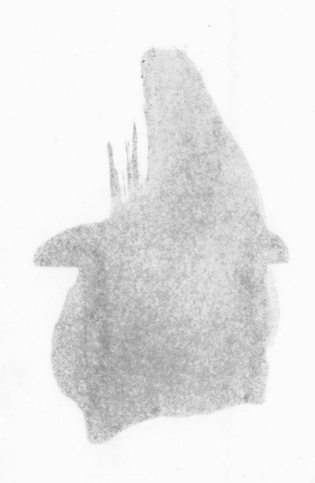